THE POWER OF ARGUMENTATION

T0353555

POZNAŃ STUDIES
IN THE PHILOSOPHY OF THE SCIENCES AND THE HUMANITIES

VOLUME 93

Poznań Studies in the Philosophy of the Sciences and the Humanities
is sponsored by SWPS

Address: dr hab. Katarzyna Paprzycka, Prof. SWPS · Department of Philosophy · SWPS
ul. Chodakowska 19/31 · 03-815 Warszawa · Poland · fax: ++48 22 517-9625
E-mail: PoznanStudies@swps.edu.pl · Website: http://PoznanStudies.swps.edu.pl

NEW TRENDS IN PHILOSOPHY

Katarzyna Paprzycka (editor-in-chief)
Department of Philosophy · SWPS
ul. Chodakowska 19/31 · 03-815 Warszawa · Poland
Katarzyna.Paprzycka@swps.edu.pl

New Trends in Philosophy is a new subseries of the *Poznań Studies in the Philosophy of the Sciences and the Humanities* book series. It publishes collections of papers that deal with new or underrepresented topics in philosophy.

Other volume in the series:

POZNAŃ STUDIES IN THE PHILOSOPHY OF THE SCIENCES AND THE HUMANITIES, VOLUME 93
NEW TRENDS IN PHILOSOPHY

THE POWER OF
ARGUMENTATION

Edited by

Enrique Suárez-Iñiguez

Amsterdam - New York, NY 2007

The paper on which this book is printed meets the requirements of "ISO 9706:1994, Information and documentation - Paper for documents - Requirements for permanence".

ISSN 0303-8157
ISBN: 978-90-420-2287-4
©Editions Rodopi B.V., Amsterdam - New York, NY 2007
Printed in The Netherlands

CONTENTS

Los seres humanos verdaderamente grandes
son siempre sencillos y buenos.
Así fueron mi Madre y mi Padre:
A ellos dedico este libro.

Truly great human beings
are always simple and good.
Thus were my Mother and my Father:
To them I dedicate this book.

PREFACE

In the days we live, days of vertiginous social, political and economic as well as cultural changes, days of spectacular scientific and technological advances, where we have seen the walls of intransigence collapse and ideologies and totalitarianism brought down, Popper's work rises with the strength that reason provides and the confidence in the best qualities of human beings. If it was a flame of hope when it first appeared, today it is an established reality. If then it led the way due to the power of argumentation, today it has gained an intellectual place.

Popper's work has transcended time and space and can therefore be considered a classic one. Translated into more than twenty five languages, both his philosophy of science and his political philosophy have been a source of controversy and criticism, of acceptance and conviction. Popper's work has stirred consciousness and raised doubts, suggested ways forward and answered questions, recovered rational tradition and given it a new expression, a suitable framework for our culture and century. Popper's work is a source of refreshing water, although not always peaceful.

His teachings are always with us. He criticized essentialism with its "what is it" and "when did it come about" kind of questions; he was against holism and exalted the importance of the individual and institutions; he denied the existence of unavoidable historical laws and the validity of prophecies in the social sciences; he rejected the presence of keys to understand history; he asserted that the theory of wiping the slate clean is dangerous and harmful, among other things, because of its uselessness; he criticized both psychoanalysis and behaviorism, Marx and Plato.

He also taught us that it does not make sense to collect information without a previous theory; that in research one does not work with subjects but with problems; that we can never demonstrate universal laws, but only refute them; that science works deductively and that we learn through conjectures and refutations and by means of gradual adjustments; that one can and should choose from among different theories the best one. He reclaimed Socrates' rationality and his concern for truth; he used Kant's postulate of autonomy of conscience as the highest principle of morals; and he emphasized that man's progress lies in his own efforts. What else could we ask of him?

Most – not all – of the contributors of this book are Popperians, some were Sir Karl's students in his famous seminar at London School of Economics and his research assistants. All have written books or papers on Popper's philosophy; one of them wrote the first introductory book to Popper's work and another was the editor of a Newsletter about the Austrian philosopher. All have their own work and now work in universities of Canada, England, Israel, Mexico, New Zealand, and the United States. So, from a well-acquainted view of Popper's philosophy, this book deals with present-day philosophical problems and offers interesting interpretations.

The first part of the book is devoted to political philosophy and the second to philosophy of science. After analyzing Popper's conception of the social, Jarvie examines if scientific institutions are open or closed, if they are just grown or are designed and reflects on the confrontation between scientific and political values. Fred Eidlin identifies Popper's critiques to Marx: the historical prophecy as the way to approach social problems; the responsibility in the downfall of democracies; his moral positivism; the proposed of revolution as the only way of building a new society and the refusal to accept criticism, both by arguments or by facts. Eidlin says that Popper does not demonstrate some of these accusations. Harold P. Sjursen criticizes Popper's interpretation of Plato's *Republic* and maintains that to argue as Popper does, that Plato's idea of justice was the bacterium that produced totalitarianism is a misreading of Plato.

From his political experience Bryan Magee explains how to apply Popper's methodology of problem solving to politics, but also affirms that it is too theoretical, too little concerned with the actual lives of individuals and that this is a serious problem in intellectual approaches. In my Chapter I maintain that behind Popper's ideas there is an educational conception that *crowns* all his political philosophy and that it is not by chance that appears just at the end of *The Open Society and Its Enemies*.

The second part on philosophy of science opens with Joseph Agassi's paper on corroboration, spurious and genuine. Agassi applies the difference between public and private aims to science. He points out that the aims of science are not always the same of scientists and relates this with success than can be spiritual or worldly. The role of science is to discover something not to corroborate theories. John Watkins analyses what he calls the Spearhead Model that emerges from Popper's evolutionary epistemology. This model deals with the relationship between the central control system of an animal and its motor system. Watkins asserts that they are genetically independent of one another and relates this to the mind-body problem.

Mario Bunge affirms that the word "rationality" stands for seven different concepts: conceptual (minimizing vagueness or imprecision); logical (striving for consistency); methodological (doubting, criticizing and demanding evidence); epistemological (empirical support); ontological (consistency with scientific knowledge); valuational (striving for worthwhile goals); practical (means-ends). Bunge maintains that Popper's rationalism is limited in the first five concepts and has "sinned" against ontological, so Popper is a half-way rationalist. Ambrosio Velasco maintains that Popper's methodological approach – situational analysis – is in fact a hermeneutic approach. It is a process of historical transformation of scientific traditions in such a way that upon the inherited background rises new problems that demand the background's transformation.

David Miller affirms that the propensity interpretation of probability is the running thread of the three volumes of *The Postcript* to *The Logic of Scientific Discovery* and that Popper relates it with indeterminism. So, Miller analyses metaphysical determinism (Landé's Blade); scientific determinism (formulated in a testable form); indeterminism, and relates all this with propensities. Miller concludes that the world is not a world run wholly by propensities but contains many chance events for which there never was any propensity of occurrence. Alan Musgrave in an attempt to put the house in order among the Popperians explains Popper's critical rationalism and his solution to the problem of induction. He disputes with both Popperians and Popper's opponents. According to Musgrave either Popper answers Hume in the way he thinks, or he, Popper, has no answer and his numerous critics on the point are right.

Most of these essays on the philosophy of Karl Popper were originally constituted a series of lectures given at the National University of Mexico and later published in Spanish (Suárez-Iñiguez 1997). Now the book has a new preface and three new chapters. One by Ambrosio Velasco especially written for the book; one by Mario Bunge that was published in *Rationality: The Critical View*, edited by Joseph Agassi and I.C. Jarvie (1987, pp. 5-15). Here it is published with kind permission of Springer Science and Business Media and of Bunge and Agassi; and a new different chapter by Eidlin. After the Congress at Mexico, the paper of Magee, Miller and Watkins were published in O'Hear (1995). Later Musgrave's paper was published as two chapters of his (1999) book with due acknowledgement to its original publication in my (1997). I thank the authors, who retained the rights in their papers, their consent to publish them here.

Popper is one of the most important philosophers of his time. As this book is such a collection of essays, I have confidence that it will catch the interest of the reader.

I wish to thank Joseph Agassi for his kind and generous help for the publication of this book, to David Miller for his friendly suggestions and to my assistant Khemvirg Puente for his help in the revision of proofs.

Enrique Suárez-Iñiguez

REFERENCES

Agassi, J. and I.C. Jarvie, eds. (1987). *Rationality: The Critical View*. The Hague: Martinus Nijhoff Publishers.

Musgrave, A. (1999). *Essays on Realism and Rationalism*. Amsterdam: Rodopi.

O'Hear, A., ed. (1995). *Karl Popper: Philosophy and Problems*. Cambridge: Cambridge University Press.

Suárez-Íñiguez, E., ed. (1997). *El Poder de los Argumentos*. México: Coordinación de Humanidades de la Universidad Nacional Autónoma de México y Grupo Editorial Miguel Ángel Porrúa.

PART I

POLITICAL PHILOSOPHY

I.C. Jarvie

THE DEVELOPMENT OF POPPER'S CONCEPTION
OF THE SOCIAL[1]

Popper is usually interpreted as first and foremost a student of the logic of the natural sciences; his work on the social sciences is treated as something of an afterthought, or perhaps an application. This paper argues to the contrary that Popper's philosophy of natural science already shifts questions of the methodology of natural science from logic and language to the design and maintenance of social institutions. He is thinking socially from the beginning.

Theses. 1. Popper's methodology of science as expounded in his 1935 book *Logik der Forschung*,[2] embodies an implicit conception of the social, a conception that was articulated, generalized and applied in his later works "The Poverty of Historicism," of 1944-1945, and *The Open Society and Its Enemies*, of 1945. From the start, Popper conceives of the social in institutionalist, reformist, piecemeal and, it would also seem, consensualist terms.[3] Treating Popper's philosophy as social all the way down to its roots in the philosophy of science, or all the way back to the

---

[1] Early versions of this paper were read to audiences at the Central European University, Prague and the University of Warwick on the 1st and 9th of June, 1994, respectively. I am grateful for the critical discussion at both those occasions, and especially to Ernest Gellner, whose thoughtful questions and trenchant writings have affected its final form. The argument sketched here was set out at greater length in a monograph *The Republic of Science*.(Jarvie 2001).
[2] The original text of *Logik der Forschung* is preserved, and translated for English-speakers, within *The Logic of Scientific Discovery* of 1959. In working with it one needs always to bear in mind Popper's own warning that every translation is an interpretation ("Reply to a Critic" 1996a, p. 326), and hence that in any translation things are always added/lost.
[3] Other "social" interpretations have been put forward by Shearmur (1985, pp. 272-282), Bartley, III (1990), Curtis (1989, pp. 77-113), Wettersten (1992). None of these, as far as I can discern, anticipates the present approach.

In: E. Suárez-Iñiguez (ed.), *The Power of Argumentation* (*Poznań Studies in the Philosophy of the Sciences and the Humanities*, vol. 93), pp. 15-29. Amsterdam/New York, NY: Rodopi, 2007.

first major publication, makes for a fruitful overall interpretation of his philosophy. Its also suggests that because most extant characterizations of his philosophy totally overlook this aspect of it they fail to connect with the most innovative aspect of his ideas.

The paper is in three parts. In the first part I extract the implicit view of the social that is to be found in *The Logic of Scientific Discovery*. In the second part I sketch how this implicit view was articulated and generalized in "The Poverty of Historicism" and *The Open Society and Its Enemies*. In the third part I develop some comments and questions toward an assessment, testing Popper's conception of the social by using it to look at the social aspect of science.

1. The Social in the Logic of Scientific Discovery

The received view of Popper's philosophy of science, as expounded by such as Salmon or Grunbaum, is that it is an argument for shifting the role of evidence in science from that of supplying positive reasons for accepting theories over to that of providing negative reasons for rejecting theories. He is said to have proposed falsifiability as the criterion for the scientific character of theories. Thus presented, Popper can be domesticated as one playing by the rules of contemporary academic philosophy of science. By contrast, my view is that Popper's position is subversive of the academic approach to the philosophical problems of science. He argues the necessity, if certain objections are to be overcome, of finding a solution to the problem of demarcation at the level of social institutions. My contention is that he saw the inadequacy of all narrowly philosophical approaches to the problems he was working on, and came to realize that if they could be solved it could only be by social technology, by decisions to reform institutions. Why he did not make this sharper and clearer in his text is not a matter I want to enter here; but I am convinced that the bold originality of his move does much to explain the woeful failure of so many in the philosophical community even to report his ideas accurately.

In contrast to the received view, I would hold that from the first *The Logic of Scientific Discovery* treats science as a social institution and conceives the problem of method as a problem of institutional reform. For the Popper of 1935 already there is no Crusonian science,[4] and

---

[4] Robinson Crusoe, alone on his island, cannot create science. Science requires social institutions.

science is not a form of personal knowledge.[5] Cooperation between persons under a régime of institutionalized rules governing procedures are necessary ingredients of science. Let me argue this by putting a few brief passages of this early work under the microscope.

Popper proposes a falsifiability criterion to demarcate empirical science: "it must be possible for an empirical scientific system to be refuted by experience" (1959, p. 41). But immediately after he proposes his criterion of falsifiability in §7 (pp. 41-42) of *The Logic of Scientific Discovery*, Popper articulates three criticisms of this view. The first criticism is that it is wrong-headed to confine science to the delivery of negative information. This he answers with the argument that a statement conveys more, the more singular statements it is likely to clash with, so negative does not mean uninformative; on the contrary, information is improbability. These claims have been endorsed by keepers of the received view, yet dissatisfaction with Popper's view as a "negative" one continues to be expressed. The second objection is that falsifiability is vulnerable to the same objections as is verifiability. This is answered with the important observation of the logical fact of the asymmetry of verification and falsification, the former being unachievable, the latter logically possible: endorsing a singular statement does not entail endorsing its generalization, but it does entail rejection of the generalization which is its negation. Accepting a statement such as "here is a black swan" does not compel us to accept "all swans are black," but it does compel us to reject, on pain of contradicting ourselves, "all swans are white." It is quite unclear whether received opinion has come to terms with this simple but decisive logical fact.

The third objection Popper raises against his own criterion of demarcation is declared to seem more serious: it is easy to evade any refutation with the aid of some *ad hoc* hypothesis which explains the refutation away. The availability of such ad hoc devices is treated as unproblematic, since one can always introduce some auxiliary hypotheses or narrow down the denotation of some terms in the refuted theory so as to exclude the refuting case. Indeed, no contradiction is involved even in such an extreme maneuver as the simple refusal to acknowledge any falsifying experience whatever. The seriousness of this third objection is that it seems to neutralize the value of the proposed falsifiability criterion. If no contradiction is involved in shielding ideas from falsifying experience then any system can be adapted to satisfy this criterion of demarcation.

---

[5] The anachronistic language is an allusion to Michael Polanyi (1958).

Popper admits the "justice" of the third objection, but goes on to say that he need not to withdraw his falsifiability proposal because he is

> . . . going to propose . . . that the empirical method shall be characterized as a method that excludes precisely those ways of evading falsification which . . . are logically admissible. According to my proposal, what characterizes the empirical method is its manner of exposing to falsification, in every conceivable way, the system to be tested. (1959, §6, p. 42)

It is notable that Popper here acknowledges the justice of a logical objection by admitting that there is no strictly or purely logical answer to it. His answer is a policy proposal: he suggests that we refuse to license ourselves to rescue hypotheses ad hoc, and choose instead his proposal that we forswear both rescuing *and ad hoc* hypotheses. He proposes that we adopt a methodological approach which refuses to countenance all ad hoc maneuvers and all waiving away of falsifying evidence.

This emphasis on policy choices greatly clarifies why Popper earlier distances himself from naturalism, the doctrine that the problems of the logic of science are part of science. Yet his approach is not conventionalist. The conventionalist ascribes to theories truth by convention; the naturalist ascribes to them truth by nature; Popper ascribes to them truth values and suggests the convention of *avoiding* ascribing truth to them by convention. A name for his approach would be methodological conventionalism – the only way in which Popper is a conventionalist.

> My criterion of demarcation will accordingly have to be regarded as a *proposal for an agreement or convention.* As to the suitability of any such convention opinions may differ; and a reasonable discussion of these questions is only possible between parties having some purpose in common. The choice of that purpose must, of course, be ultimately a matter of decision . . . (1959, §4, p. 37)[6]

What Popper proposes in this short passage is remarkable: that the problem of demarcation cannot be solved satisfactorily within the logic of statements. It can only be solved by an agreement or a convention. The boundaries of science are like national boundaries, they are manmade

---

[6] The passage ends with the phrase, "going beyond rational argument." This parting sally of Popper's suggesting that decision goes beyond rational argument is consistent with the view he articulated in *The Open Society.* It has been the centre of some controversy, with his disciple and critic Bartley ending up labelling Popper "fideistic", in the second edition of his *The Retreat to Commitment* (1984, p. 104, p. 215n). This seems to me a fundamental misunderstanding of a simple point.

social institutions. Popper is a one man Boundary Commission offering a proposal. Opinion may differ on the suitability of a suggested boundary, which will then be open to discussion. In order for this discussion to be fruitful, the parties to it must have some purpose in common. Thus, Popper is suggesting that his demarcation criterion is a proposal for institutional reform: reform guided by the aim of maximizing the impact of experience on hypotheses. The proposed boundary, the guiding aim, and whether the boundary proposal subserves the aim are the obvious points for discussion.

It is easy to overlook the fact that we have here Popper's first published discussion of what he years later defended as piecemeal social engineering. We have a problem, namely, setting the boundaries of science. And we have various boundary proposals, many of which take it for granted that the problem is not institutional but natural. Popper argues that none of these proposals, including his own, will withstand logical scrutiny. The demarcation problem is not solvable within logic and language. This in turn is because science is more than simply a set of statements; it is a set of statements produced in, and governed by, a social context of practices, traditions, institutions, and it is only in that social context that they can become scientific. Outside that social context the self-same statements need not be scientific. Hence, whereas the naturalistic view is not viable, the conventionalist view is viable but, Popper argues, it is objectionable as too defensive.

Popper's analysis has shown him and us that the demarcation problem is social, hence, that its solution is social, i.e., social reform, and that in considering the reform of social institutions the very first question to be addressed is, what is the aim common to the reformers? He proposes that the aim of science is to learn from experience, to use, as he says, experience as a method. If experience is to be used as a method, then allowing theories to be protected from falsification by experience is inadvisable. Conventionalists propose ways to avoid refutation, making it possible for any statements to be scientific. Popper offers the contrary convention of welcoming refutation, thus narrowing the range of statements which can be scientific. Yet, he does not explicitly go further and point up the fact that he has treated science as a social activity. He leaves it to us to notice that an agreement or convention is an arrangement between people who share an aim, and to foster that aim they undertake to follow certain agreed-upon rules. In short, they form a social institution.[7] In this way Popper effected a revolution in the philosophy of science: for him science is not a naturally bounded set of statements, but a set of activities the aim of which is the production of statements under the régime of the method of experience. Only statements so produced are candidates for scientific status.

So when we think of scientific method as Popper envisages it, we would be in error-to ask the question whether he is describing it or proposing something normative. Under the pervasive influence of Kant he is, rather, proposing something constitutive: the agreement or convention he wants to put in place will create a particular design of social institution, a pre-condition for realizing the aim of science. Science is envisaged as consensual, because cooperation in operating its rules is necessary; as goal-directed, since it has been constructed (or reconstructed) so as to achieve an aim; it is reformable because both its aim and the effectiveness of the means to the aim can be rationally discussed and proposals brought forward for alteration; and above all it is an institution, a permanent structure to coordinate and direct human activities. It is this institutional and reformist view where Popper shows his methodological conventionalism:

> what is to be called "science" and who is to be called a "scientist" must always remain a matter of convention or decision. (1959, p. 52)

Long before he wrote *The Logic of Scientific Discovery* Popper had worked in social work and in education and so had much opportunity to reflect on the nature of social things. In thinking about science Popper reveals that he thought of it as a social activity, engaged in by like-minded persons, whose like-mindedness was not their agreement in judgment but their agreement in aim and method. He shows an acute awareness that sociology is about the institutional frameworks within which both individuality and creativity can be nurtured and disciplined. He resembles here a Kantian legislator, or perhaps a constitutional delegate proposing basic law within which social life can be continued on an agreed basis, basic law that is not written in stone, but put forward for certain purposes, and, even after implementation, is open to still further debate and reform. Basic law is not easy to alter, but it must be open to discussion and reform, nonetheless.

---

[7] Admittedly this account is an over-simplification. Social rule-following is not like a meeting agreeing to follow Roberts' *Rules of Order*. Institutions and the apparent rule following behaviour that constitutes them are mostly forms of spontaneous social order ("just grown," see below), for which the classical example is the market. On this see Shearmur (1985) and especially Curtis (1989).

Before continuing with the articulation and generalizing of these ideas in part II, I want to make two critical comments that will be developed in part III. Although in *The Logic of Scientific Discovery* Popper clearly treats the demarcation of science as a social rather than a psychological matter, he says little or nothing about what we might term the internal social structure of scientific institutions. If what constitutes science is submission to a methodological regime, there are many questions to be answered about how that regime is set up, administered, maintained, and reformed. These questions might take one in the first instance to the history of science, and in the second instance to the sociology of science. In *The Logic of Scientific Discovery* Popper is more or less silent on both, and, moreover, he does not suggest either one as a research programme that might throw some light on what he describes as "the choice of methods" (p. 49):

> our manner of dealing with scientific systems: by what we do with them and what we do to them. Thus I shall try to establish the rules, or if you will the norms, by which the scientist is guided when he is engaged in research or in discovery, in the sense here understood. (1959, p. 50)

This reticence is puzzling, since it blocks him from pursuing some of the ideas advanced in *The Logic of Scientific Discovery*. This reticence of his also goes some way to explain why the debates that have surrounded Popper's book have rarely focused on the issues raised here and hence have seldom debated the proposed aim and the regime of rules proposed to foster the aim.[8]

The second comment is about the aim of science. In the early part of *The Logic of Scientific Discovery* Popper indicates that science aims at the presentation and testing of falsifiable statements. This much is straightforward enough. But immediately the further question arises, what is the aim of aiming at falsifiable statements? Popper's answers are internal, that is, he has explained the virtues of systems of falsifiable as opposed to verifiable or merely conventional statements. Thus a falsifiable system is an empirical system, one where our knowledge is subordinated to experience. The values and worth of the whole enterprise Popper takes for granted. So we get no argued alternative to, for example, Bacon's view that the aim of science is to understand nature in order to get power over her. Indeed in 1935 we find Popper explicitly repudiating an evolutionary approach to science as inadequate (he seems to equate it

---

[8] Two notable exceptions are Johansson (1975) and Blaug (1980).

with instrumentalism, see 1959, p. 278), and closing his book with a powerful passage intimating that science is an end in itself:

> science never pursues the illusory aim of making its answers final, or even probable. Its advance is, rather, towards the infinite yet attainable aim of ever discovering new, deeper, and more general problems, and of subjecting its ever tentative answers to ever renewed and more rigorous tests. (1959, p. 281)

2. The Elaboration and Generalization of Popper's Conception of the Social

Popper's two principal works on the philosophy of society and of politics were completed in the ten years after *The Logic of Scientific Discovery*, "The Poverty of Historicism" (1944-45) and *The Open Society and Its Enemies* (1945).

"The Poverty of Historicism" explicitly portrays social institutions as tools or means for the achievement of social aims, but adds a significant qualifier. Only some institutions have been consciously designed for a purpose; the majority have just "grown" Designed institutions have ready-made criteria for assessment of their performance. Institutions that have just grown are more problematic: by what standards should they be judged and, since they are an inheritance, what control are we entitled to exercise over them? The question of democratic control over, and reform of, institutions that have just grown is not explicitly addressed in "The Poverty of Historicism," but a related matter is. I refer to two sorts of reform policies, the distinction between piecemeal and utopian social engineering. In his discussion, Popper treats all institutions equally: they are hypotheses the efficacy of which we should test, and he argues that if we reform them we must proceed piecemeal because large-scale social reform is untestable and hence self-defeating.

Two other innovations introduced in "The Poverty of Historicism" were a methodological rule for social explanation and a stress on the unintended consequences of institutional innovation and reform. The rule of methodological individualism says that to avoid essentialism towards the social entities used in our explanatory models, we do better to think of our models in descriptive or nominalist terms, "*in terms of individuals,* of their attitudes, expectation, relations, etc."[9] Although some institutions

---

[9] In the book version of "The Poverty of Historicism" of (1957) this is at p. 136. In the original articles (*Economica* **12**, 1945) it is at p. 80, with only minor syntactic changes.

are purposely designed both they and grown institutions are part of a large network that interacts in ways too complex to predict, even in the short term. Thus institutions require maintenance, reform and sometimes dismantling, depending on how they are performing. Much of this is generalization of what was explicit but particular in *The Logic of Scientific Discovery*.[10] Science was there assessed against an aim, and its deficiencies remedied with proposals for a reformed set of values. There was no essentialist appeal to the ethos of science or anything similar; the proposed methodological rules are to be evaluated by their consequences, and the rules are intended to create an institutional situation that will promote a certain social outcome.

The Open Society and Its Enemies is much longer than "The Poverty of Historicism" and much richer in its discussions of these matters. What are described as the social aspects of scientific method are explicitly addressed, and a strong parallel is drawn between the community of scientists united in cooperative rational pursuit of the truth, and the enlightened approach to the reform of society and social institutions in general. The specialized rationality of scientific institutions is treated as a model for democratic politics: institutional structures that enjoin open-mindedness and the critical attitude are recommended, the aim being applied or useful knowledge about the way to achieve social aims.

Popper's conception of social institutions is quite a bit more explicit and developed in *The Open Society and Its Enemies*. Hypothetical knowledge itself is seen as a social institution and institutions embody hypothetical knowledge. This is because institutions are experiments at reform or change to be implemented and tested in practice. Society accumulates knowledge in its institutions, society is thus a knowledge-accumulating entity. This boldly generalizes the idea that scientific knowledge is social and that scientific institutions need to be constructed in such a way as to maximize their potential to foster knowledge.

Two further innovations in *The Open Society and Its Enemies* are extensions of *The Logic of Scientific Discovery*. First is Popper's general characterization of societies as open or closed and his historical idealization of the direction of social change from concrete or face-to-face societies towards abstract and anonymous ones. Second is the emphasis on the autonomy of sociology and the recommendation that

---

[10] The argument that Popper's methodology is redundant and this his other idea of the logic of the situation is sufficient to make sense of the progress and rationality of science was made by Hattiangadi (1978, pp. 345-365 and 1979, pp. 49-76). This seems also to have influenced Curtis (1989).

social   problems   should   be   approached   sociologically,   not
psychologically.

Science as a social activity is the model for the open society,
especially as the prototype of science is Socratic dialogue. This is a
social, intellectual and moral ideal in *The Open Society*, where it is
strongly contrasted with the hierarchy, dogmatism and irrationality of the
closed or tribal society. Because there is no science without social
institutions, Robinson Crusoe's lonely scientific effort is not science.[11]

The anti-psychologism of *The Open Society* is independent, but
dovetails very nicely with the anti-psychologism of *The Logic of
Scientific Discovery*. In that latter book intersubjectivity is substituted for
detachment, institutions for mental preparation, and methodology for
epistemology, where methodology is agreed rather than natural rules. It is
remarkable that Popper, steeped in psychology as he was in the 1920s, is
one of the sharpest critics of any tendency to psychologise social
matters.[12]

3. Some Comments and Questions

We find that Popper's conception of the social is that there is no science
without scientific institutions; there is no objectivity to science without
its institutions; and the empirical method is necessarily piecemeal
because it is impossible to test and try to eliminate all errors in one go.
Science is a model for how we use reason to learn, and hence a model for
the moral unity of mankind. It is only by cooperating in social institutions, while reducing the traditional (social and cultural) barriers to
cooperation, that knowledge is obtainable at all. We are thus mutually
interdependent and our mutuality has to do with a recognition that we are
all in the same boat and in need of one another to accomplish such
projects as science.

Why, then, did not Popper more clearly proclaim and develop his turn
towards the social embodied in *The Logic of Scientific Discovery*? I raise
the question, but do not wish to speculate upon an answer.

---

[11] *The Open Society and Its Enemies*, Popper 1966b, Chapter 23, pp. 218-220. Objectivity
is shown to depend on cooperation; the empirical method itself is always piecemeal and
can work only if we are each checked by others; individual bias and prejudice cannot be
eliminated and the institutions are there to compensate for it. Using other words, virtually
the same argument opens Polanyi (1962, pp. 54-73).
[12] Hacohen (1993) makes a convincing case that this is not unconnected with the
Methodenstreit between German and Austrian economics.

There are other specific questions for the answers to which one searches his work in vain. Are scientific institutions grown or designed; open or closed; abstract or concrete; and do they change piecemeal or holistically?

The key question here is whether science as institution is grown or designed. Let me distinguish science in general from science in particular. Science with a capital *S*, the world-wide invisible college, has, it seems to me, just grown. (This is consistent with many of the elements of the aggregate, such as international scientific societies, being designed.) Science in its particular institutional embodiments, from the Royal Society to the Manhattan Project, is largely contained within designed institutions. Only designed institutions have aims; although grown institutions can have aims attributed to them. Unless science is designed, there is something puzzling about discussing its aim. Furthermore, both designed and grown institutions have functions as well as aims, and the functions may not be congruent with the aims. The Royal Society was not founded with any anticipation of the function it would perform in setting world-wide standards for Science, or even of being a world-wide model for scientific societies. The creators of the Manhattan Project to develop the atomic bomb did not aim to create a model for all future "big science," still less for its colossal and normative impact on science proper.

Are methodological rules an incidental part of science or constitutive, as Popper's writing seems to suggest? And, more important, how did the extant body of rules come into being? And, still more important, how are they to be reformed? In which forum, by whom, and how are any changes promulgated? Despite his institutionalism, Popper does not follow through on the consequences of his ideas in their social application; he persisted with an amorphous model of science. When we are invited to consider a model for the learning society in *The Open Society*, the invitation is to admire the critical rationalism of Socrates; and it is quite apparent that this is also Popper's model for science. Socrates conducted philosophy with a small circle of friends and hangers on, in a group that appears to have had no internal structure at all.[13] Popper must have known very well that as a model for the actual practice of science this was an idealization – to say the least. The debates surrounding the ideas of Einstein and Bohr, being largely theoretical, may be thought to have constituted a transnational Socratic seminar. The social ramifications of

---

[13] Incidentally, but relevantly, the aim of the group was to care for one's soul, not to gain knowledge of the world.

that seminar bore no resemblance to the practice of Socrates – they included the Manhattan Project. Laboratory science, as pursued in industry, follows the model of the Manhattan Project: in it large teams, large resources and division of labor are coordinated in institutions very different from Socratic seminars. This kind of science is conducted in institutions with clear internal structure, such as hierarchy, specialization, scale, compartmentalization, division of labor and barriers to entry. The transnational institution of Science as such, the widest construal of Robert Boyle's invisible college, is by interesting contrast much more diffuse, organized more like the Internet than like a hierarchy. Be it noted that in their institutionalization, neither concrete science nor science in its world-wide sense resembles Popper's Socratic model.[14]

Popper's devotion to Socratic face-to-face critical discussion in small undifferentiated groups is longstanding; it is a model first for science (*The Logic of Scientific Discovery*, 1935), then for philosophy, then for the open society at large (*The Open Society*, 1945). Given that he is a bold critic of other social institutions, such as the market or the educational system, his not examining the actual workings and actual shortcomings of the institutions of science is difficult to understand. From the Royal Society to the German chemical industry there is a real history and a real sociology of the institutional embodiments of science, containing no doubt many dangers and mistakes that we would do well to learn from and avoid. And the most ticklish point is what one might call the politics of science: the formation of schools and parties and client-relationships and laboratory cultures and Internet groups that can be only partly explained by divisions over what is the truth of the matter and how we should advance towards it.

Popper had a golden opportunity to discuss the actual institutions of science when he and Kuhn were brought together for a debate in 1965. Instead, he made the jarring claim (Popper 1970, pp. 51-58) that Kuhn was his "most interesting" critic, one who understood him better than most, and had seen clearly and named a kind of scientific activity – normal science – that was inimical to Popper's idealization. Yet Kuhn's sociology was meager and he said little about the politics of science and did not analyse major scientific institutions. Fair enough; Kuhn's sociological ideas are thin and rather quickly run off into the sands of psychologistic reductionism.[15] Not so Popper, who we have seen to have a quite distinct sociology, with a theory of individuals and a theory of

---

[14] Nor, I think, is it quite as republican as depicted in Polanyi (1962).
[15] I am thinking of Kuhn's talk of Gestalt switches, the weight of opinion, and the like.

institutions and traditions and a theory of how individuals reform institutions – all of which could have yielded a rich discussion of Kuhn's historical materials about teaching, socialization, textbooks and leadership. These remarks concerning Popper's omissions hardly amount to criticism; they are more expressions of disappointment at consequences not thought through.

To illustrate, consider one very fruitful idea. Poppers says in *The Open Society* that institutions should always be framed on the assumption that anti-democratic tendencies are ubiquitous, whether latent or overt, in leaders and in followers. The application of this warning to science could explain why science has so proliferated institutions – proliferation is one way to end run anti-democratic tendencies in extant institutions – and also why science is so riven with politics, which may be connected as much with populist anti-democratic tendencies as with divisions over the truth. Politics and proliferation are means to mediate the allocation of power. In idealized science the only power should be the power of truth – the power of theory, the power of a critical argument. But as soon as there is an organized structure there are other kinds of power that do not harmoniously correlate with truth and argument. For example, there is power in establishing the constitution of the organization; power in administering it; power in setting its agenda; all of these raise question difficult to settle by appeals to the power of current theories or of arguments, especially as the author of a theory or argument may espouse views on the constitution, administration and agenda of the institution that are not congruent with the aims of science.

Thus the client system – you take my students, I take yours – which is particularly important in science recruitment, is not seen by its operators as a method of perpetuating and entrenching mediocrity; when that happens it is treated as an unintended consequence; it is rather seen as a piece of rational ordering in an otherwise haphazard process. The bosses in the client system often do their utmost to internalize very high scientific standards in their students, even if their methods, such as ruthless bullying, might seem destructive.

If patronage, anti-democratic tendencies, power and politics are problems for scientific institutions, then Popper's own ideas suggest some thought needs to be given to their design and manning[16] and in particular to their mechanisms for self-reform. Otherwise it is unclear where the methodological rules of the kind he puts forward find their

---

[16] "Institutions are like fortresses. They must be well designed and manned," *The Open Society and Its Enemies*, Popper 1966a, Chapter 7 (III), p. 126.

home, in what forum and under what rules they can be debated, and how if at all the amended or supplemented rules that emerge from the debate can be promulgated.

To this it might be objected that Popper thought that science aims at ever more falsifiable hypotheses, ever deeper problems, which in his later philosophy he was content to call the truth. Ought not scientific institution be designed and judged solely against their fostering of that aim? My answer is that there is perhaps a problem with the pursuit of truth as an organizing principle for societies or social institutions: truth can undermine organization, create problems for community building, for institutional cohesion.[17] One reason is that truth is indifferent to all other authorities, such as tradition, the law, custom, seniority, training, or hierarchy. Institutions and social relations dependent on any of these are vulnerable to truth. Furthermore, what the truth is on any matter, on Popper's view of science, is open to constant dispute. Thus, there is at best the possibility of limited and temporary consensus developing around truth, not to mention the fact that truths are subject to interpretation, especially in their application. For these reasons, the scientific pursuit of the truth is a most unpromising prospect around which to organize society and its institutions. Truth and social institutions seem to be inimical to one another; either truth undercuts the very structure of the institution itself; or the institution develops structure and practices that inhibit the pursuit of truth. The sociology of how these tendencies are struggled against and partially overcome is a problem Popper's methodological revolution directs us to explore. Popper, as I have tried to show, was always a deep and penetrating thinker about society and social institutions; thus it is a matter of bitter regret that the opportunity of discussing these difficulties with him is now lost.

York University
Department of Philosophy
S428 Ross Building
4700 Keele Street
Toronto, Ontario M3J 1P3
Canada
e-mail: jarvie@yorku.ca

---

[17] This argument has been made by Gellner in several places, most succinctly in *Conditions of Liberty* (1994, pp. 31-32).

REFERENCES

Bartley, W.W. III (1984). *The Retreat to Commitment*. La Salle, IL: Open Court.

Bartley, W.W. III (1990). *Unfathomed Knowledge, Unmeasured Wealth: On Universities and the Wealth of Nations*. La Salle, IL: Open Court.

Blaug, M. (1980). *The Methodology of Economics*. Cambridge.

Curtis, R. (1989). Institutional Individualism and the Emergence of Scientific Rationality. *The History and Philosophy of Science* **20**, 77-113.

Gellner, E. (1994). *Conditions of Liberty*. London: Hamish Hamilton.

Hacohen, M. (1993). The Making of the Open Society: Karl Popper, Philosophy and Politics in Interwar Vienna. Ph.D. dissertation: Columbia University.

Hattiangadi, J.N. (1978). The Structure of Problems I. *Philosophy of the Social Sciences* **8**, 345-365.

Hattiangadi, J.N. (1979). The Structure of Problems II. *Philosophy of the Social Sciences* **9**, 49-76.

Hattiangadi, J.N (1983). A Methodology without Methodological Rules. In: R.S. Cohen and M.W. Wartofsky (eds.), *Language, Logic, and Method*, pp. 103-151. Dordrecht: Reidel.

Jarvie, I.C. (2001). *The Republic of Science: The Emergence of Popper's Social View of Science 1935-1945*. Amsterdam: Rodopi.

Johansson, I. (1975). *A Critique of Karl Popper's Methodology* (Stockholm: Academiförlaget).

Polanyi, M. (1962). Personal Knowledge. London: Routledge and Kegan Paul.

Polanyi, M. (1962). The Republic of Science. *Minerva* **1**, 54-73.

Popper, K.R. (1934). *Logik der Forschung*. Vienna: Springer. English translation in Popper (1959).

Popper, K.R. (1957). *The Poverty of Historicism*. London: Routledge and Kegan Paul. Originally published in *Economica* **11** (1944), 86-103; **12** (1945), 69-89.

Popper, K.R. (1959). *The Logic of Scientific Discovery*. London: Hutchinson (English translation of Popper (1934), with new preface, footnotes and appendices; 3rd ed. 1972).

Popper, K.R. (1966a). *The Open Society and Its Enemies*. vol I. 4th ed. Princeton, NJ: Princeton University Press.

Popper, K.R. (1966b). *The Open Society and Its Enemies*, vol. II. *The High Tide of Prophesy*. Princeton, NJ: Princeton University Press.

Popper, K.R. (1970). Normal Science and Its Dangers. In: I. Lakatos and A. Musgrave (eds.), *Criticism and the Growth of Knowledge*, pp. 51-58. Cambridge: Cambridge University Press.

Shearmur, J. (1985). Epistemology Socialized? *Etc* **42**, 272-282.

Wettersten, J.R. (1992). *The Roots of Critical Rationalism*. Amsterdam: Rodopi.

Fred Eidlin

CALL TO DESTINY OR CALL TO ACTION:
MARX, POPPER, AND HISTORY

ABSTRACT. Popper is in substantial agreement with both Marx's ideals and theoretical achievements. Nevertheless, he accuses Marx of having misled scores of intelligent people into believing that historical prophesy is the scientific way of approaching social problems. While Marx's writings do not on balance put forward a mechanical and simplistic view of history according to which all societies were predestined to go through a single, inexorable sequence of stages, several features of these writings encourage such an interpretation. The fourth section identifies problems in Popper's critique of Marxism: his attribution of causality to Marx's ideas may itself have a historicist flavor. He does not convincingly demonstrate the responsibility of Marxism for the downfall of democracy and rise of totalitarianism. Doctrines which Popper identifies as pernicious can be and often have been tamed and put to good use in practice. It is unclear that Marx was in fact a moral positivist as Popper claims. Historicism, by Popper's own testimony pervades even the writings of liberal political thinkers. The final section zeroes on those of Popper's criticisms of Marx that hit the bulls eye: a) their deceptive way of selling revolution as the only way to bring about real change in society, excluding any concern for social technology; and b) those features that make them resistant to correction by experience and criticism.

1. Introduction

Karl Marx's views on both history and science are central to his social and political theory. On this point, critics and defenders agree. For Marxists the belief that Marx discovered the key to history provides assurance of the rightness of their political action. Marx's critics, on the other hand, have argued that flaws in Marx's theory of history fatally undermine his social and political theory, and are sufficient grounds for rejecting it. What is at stake, as Ellen Meiksins-Wood puts it, is

> not simply the value of Marxism as a theory of history and its alleged inability to account for the variety of historical patterns on display in the world, but also the viability of the socialist project. Since Marxism was

In: E. Suárez-Iñiguez (ed.), *The Power of Argumentation* (*Poznań Studies in the Philosophy of the Sciences and the Humanities*, vol. 93), pp. 31-54. Amsterdam/New York, NY: Rodopi, 2007.

so clearly wrong about the unilinear course of history, surely it was equally wrong about the inevitability – indeed the possibility – of socialism.

Defenders of Marxism agree about the importance of the issue, if not in their conclusions about it. They recognize that Marxism is indefensible if it really does rest on the kind of "mechanical and simplistic view of history according to which all societies were predestined to go through a single, inexorable sequence of stages" (1984, p. 95). Critics, on the other hand, have argued that Marxism without such a theory of history, has "lost hold of its lifeline," and is dead (p. 96).

Much also depends on the scientific status of Marx's theory of history. For Marxists, it is precisely the scientific status of Marxism that sets it apart. It distinguishes Marx's theory from utopian varieties of socialism. It also sets Marxism apart from speculative philosophical systems that merely interpret the world, when the point is to change it. Marxists claim, using such arguments, that Marxism is not just another ideology or philosophy, but a true, scientific explanation of the dynamics and evolution of society. As such, Marxist theory can serve as a reliable guide to political action. Critics, on the other hand, frequently argue that serious flaws in Marx's views on science undermine his entire theoretical system, and are grounds for rejecting it.

Karl Popper's critique of Marx's theory is of special interest since it focuses on precisely these two crucial aspects – history and science. Marxists, non-marxist specialists on Marx's thought, and anti-marxists alike acknowledge the importance of this critique. Sir Isaiah Berlin, calls it "the most scrupulous and formidable criticism of the philosophical and historical doctrines of Marxism by any living writer" (1963, p. 287). Marxist Maurice Cornforth argues that "of all the critics, . . . Popper is the one to answer." This is, not only because "he is he perhaps the most eminent of our contemporary critics, and not only does he present his case with great ability and force, but because . . . the points he makes against Marxism include practically all the main points against it which carry most weight in contemporary debate" (1968, p. 5).

Despite its importance and celebrity (or notoriety), however, Popper's critique of Marxism has not received much careful scrutiny.[1] Neither Marx's critics nor defenders have sufficiently thought through the implications for Marxism of this critique. Marxists usually malign or

---

[1] For discussion of the reception of Popper's critique, and this odd coincidence of importance and celebrity along with lack of serious critical discussion see, e.g., Suchting (1985, pp. 147-149) and Jarvie (1982, p. 83, p. 103).

ignore Popper's writings on Marx. And anti-marxists like to cite Popper as support for their claims that Marxism is pernicious and intellectually bankrupt. Marxists, in their anger, have not often been inclined to look beyond the flaws in Popper's critique. They have thus usually failed to recognize its value. At the same time, anti-marxists, including many admirers of Popper's social and political philosophy, usually do not realize how much Popper remained a Marxist. Nor are they usually aware of how troublesome certain aspects of Popper's writings ought to be for conservatives and right-wing liberals.

This article strives to clarify and assess Popper's critique of Marx's alleged historicism. Its main thesis is that Popper's charges against Marx are poorly focused. This makes it easy for Marxists to dismiss Popper's critique without grasping its incisive and important insights. This lack of focus also fosters an impression among anti-marxists that Popper has simply dealt a death blow to Marxist theory, making any serious confrontation with it unnecessary.

The first section shows how, despite the serious criticism contained in *The Open Society and Its Enemies* and *The Poverty of Historicism*, Popper expresses considerable agreement with Marx. He also shows great sympathy for Marx's ideals and theoretical achievements. When one tallies up the praise and criticism, quite a lot is left of Marxism. Indeed, it is perhaps as much as many who today call themselves Marxists would want to defend.

The second section clarifies Popper's main charge against Marx. This is that Marx was an inadvertent intellectual Pied Piper, who misled "scores of intelligent people into believing that historical prophesy is the scientific way of approaching social problems . . . " This, Popper contends, contributed to the devastating influence of the historicist method of thought within the ranks of those who wish to advance the cause of the open society.

The third section inquires into whether Marx really held the views Popper ascribes to him. There is no simple answer to this question. Marx's defenders may be right in arguing that Marx did not put forward a "mechanical and simplistic view of history according to which all societies were predestined to go through a single, inexorable sequence of stages." Yet, several features of Marx's writings, including his own understanding of them, strongly support such an interpretation.

The fourth section explores several problems in Popper's critique of Marx. It argues: a) that Popper's attribution of historical causality to Marx's ideas may itself be historicist; b) that Popper does not establish the responsibility of Marxism for the downfall of democracy and rise of totalitarianism; c) that doctrines Popper characterizes as pernicious can be and often have been tamed in actual practice; d) that Popper does not establish that Marx was in fact a moral positivist; e) that historicism, by Popper's own testimony is rampant, even among the ranks of liberal political thinkers, further weakening the link between historicism and totalitarianism.

The final section zeroes in on Popper's most important criticisms of Marx: a) that Marx's means of selling revolution as the only way to bring about real change in society are deceptive, and b) that there are features of Marxist theory, which give it a hardy resistance to correction by experience and criticism.

2. Marx and Popper's Theory of Criticism: Isn't Popper the Best Marxist?

Marxists and anti-marxists alike usually believe that Popper's aim in writing *The Open Society and Its Enemies* was to destroy and reject Marx's work. This is largely due to the common understanding of criticism as synonymous with destruction and rejection. Following Popper's fallibilist, anti-justificationist theory of knowledge, however, to criticize a theory does not mean to raze it to the ground and begin anew, but rather to improve it. Worthiness of criticism is, for Popper, a compliment. Knowledge grows through the singling out of bold theories that are rich in content. Identification of errors, even of confusions and obfuscation is a first step toward their improvement. This critical approach does not demolish a theory. Rather, it takes over what remains after error elimination.

In light of this fallibilist view of criticism, it might be more appropriate to characterize *The Open Society and Its Enemies* as a guide to improvement of Marxism, rather than an attempt to destroy it. "Marx was a rationalist," Popper writes. "With Socrates, and with Kant, he believed in human reason as the basis of the unity of mankind" (1966b, p. 224). And further, "There can be no doubt of the humanitarian impulse of Marxism. Moreover, in contrast to the Hegelians of the right-wing, Marx made an honest attempt to apply rational methods to the most urgent problems of social life. The value of this attempt is unimpaired by the fact that it was . . . largely unsuccessful. Science progresses through trial *and* error. Marx tried, and although he erred in his main doctrines, he did not try in vain. He opened and sharpened our eyes in many ways. A return to pre-Marxian social science is inconceivable. All modern writers are indebted to Marx, even if they do not know it" (1966b, pp. 81-82).

According to Popper, "Marx saw his specific mission in the freeing of socialism from its sentimental, moralist, and visionary background." He wanted to develop socialism from its utopian stage to its scientific stage (1966b, p. 83).

> His open-mindedness, his sense of facts, his distrust of verbiage, especially of moralizing verbiage, made him one of the world's most influential fighters against hypocrisy and phariseeism . . . His main talents being theoretical, he devoted immense labor to forging what he believed to be scientific weapons for the fight to improve the lot of the vast majority of men. His sincerity in searching for truth and his intellectual honesty distinguish him, I believe, from many of his followers . . . Marx's interest in social science and social philosophy was fundamentally a practical interest. He saw in knowledge a means of promoting the progress of man. (1966b, p. 82)

"Marx himself would have agreed with such a practical approach to the criticism of his method," Popper writes. In other words, Marx would have approved of asking of

> whether it is a fruitful method or a poor one, that is, whether or not it is capable of furthering the task of science . . . for he was one of the first philosophers to develop the views which later were called 'pragmatism' . . . Science, he taught, should yield practical results. Always look at the fruits, the practical consequences of a theory! They tell something even of its scientific structure. A philosophy or a science that does not yield practical results merely interprets the world we live in; but it can and it should do more; it should change the world. (Popper 1966b, p. 84)

What if Marx could come back to life and assess what had happened since his death, in politics, in the interpretation of his ideas, and in man's understanding of science? He might well consider Popper his best disciple (Eidlin 1983, pp. 18-20). The scientific status of his work was crucial to Marx. "Every opinion based on scientific criticism," he wrote, "I welcome" (1967, p. 11).

Popper is also in almost full agreement with the moral thrust of Marx's views. "Marx lived . . . in a period of the most shameless exploitation and cruel exploitation," he writes. "This shameless exploitation was cynically defended by hypocritical apologists who appealed to the principle of human freedom, to the right of man to determine his own fate, and to enter freely into any contract he considers favorable to his interests)."

Marx's burning protest against the crimes of unrestrained capitalism which were then tolerated, and sometimes even defended, not only by professional economists but also by churchmen, will secure him forever a place among the liberators of mankind" (1966b, p. 122). "Marx showed," Popper writes, "that a social system can as such be unjust; that if the system is bad, then all the righteousness of the individuals who profit from it is a mere sham righteousness, is mere hypocrisy" (1966b, p. 211).

For our responsibility extends to the system, to the institutions which we allow to persist. It is this moral radicalism of Marx which explains his influence; and that is a hopeful fact in itself. This moral radicalism is still alive. It is our task to keep it alive (1966b, p. 199, p. 211).

Such appreciation pervades Volume II of *The Open Society and Its Enemies* from cover to cover.[2] For example, Popper calls Marx's attempts to use the "logic of the class situation" to explain the working of the institutions of the industrial system admirable. Despite certain exaggerations and the neglect of some important aspects of the situation, he writes, it is admirable as a sociological analysis of that stage of the industrial system which Marx has mainly in mind: the system of "unrestrained capitalism" of one hundred years ago (1966b, p. 117). Popper also credits Marx with having improved upon John Stuart Mill by criticizing Mill's methodological psychologism. In doing so, Popper writes, "he opened the way to the more penetrating conception of a specific realm of sociological laws, and of a sociology which was at least partly autonomous" (1966b, p. 91).

To be sure, Popper criticizes Marx's view of the impotence of all politics and of "mere formal freedom." Yet, he takes care to stress that "even his mistaken theories are proof of his keen sociological insight into the conditions of his own time, and of his invincible humanitarianism and sense of justice" (1966b, p. 121).

There are, Popper asserts, two different aspects of Marx's historical materialism.The first is historicism, which he thinks we must reject. The second is economism (or materialism). This is "the belief that the economic organization of society, the organization of our exchange of matter with nature, is fundamental for all social institutions and especially for their historical development." Popper considers economicism quite sound, provided the term fundamental is taken "in an

---

[2] Popper's views of Marx's *personal motives* were later modified after he read Leopold Schwarzschild's book on Marx. In Appendix II, added to vol. II of *The Open Society* in 1965, Popper writes: ". . . though [Schwarzschild's] book may not always be fair, it contains documentary evidence . . . which shows that Marx was less of a humanitarian, and less of a lover of freedom, than he is made to appear in my book" (1983, p. 396).

ordinary vague sense, not laying too much stress upon it" (1966b, p. 106). Moreover, even in criticizing Marx for having taken the term fundamental too seriously (1966b, p. 107), Popper excuses him for having fallen victim to the prejudices of his times concerning the nature of science. In Marx's times, he points out, people generally believed that science implies determinism and that the aim of science is to predict the future.

Popper aims his main criticism at historicism. Moreover, while calling Marx a "false prophet," he contends that Marx saw himself, not as a prophet, but as a scientist. Furthermore,

> [a] closer view of Marx's successes shows that *it was nowhere his historicist method which led him to success, but always the methods of institutional analysis* . . . [C]ompared with *Marx's own high standards, Popper asserts, the more sweeping prophecies are on a rather low intellectual level.* (Popper 1966b, p. 197, emphasis supplied)

3. Popper's Charges against Marx

Such lavish praise makes it all the more interesting to look closely at the serious charges Popper levels against Marx. To be sure, he does not characterize Marx as a totalitarian party politician as he does Plato (Popper 1966a, p. 169). Rather, he portrays Marx as an advocate of the open society led astray by his historicist methodology. He certainly does not blame Marx for making mistakes. After all, the fallibility of all human knowledge is at the heart of Popper's theory of knowledge. There is nothing wrong with being wrong. Nevertheless, Popper contends that Marx's ideas contributed significantly to the rise of totalitarianism.

Although he never states outright that Marx was responsible for the horrors of Stalinism, the implication is present. Not only does Popper represent Marx's ideas as part of the intellectual legacy of Stalinism. He also contends that these ideas misled and thus weakened the forces that could and should have opposed Stalinism and Hitlerism more effectively. Popper's main charge against Marx is that he inadvertently acted as a kind of intellectual Pied Piper. He misled "scores of intelligent people into believing that historical prophesy is the scientific way of approaching social problems." In this, he contributed to the devastating influence of the historicist method of thought within the ranks of those who wish to advance the cause of the open society (Popper 1966b, p. 82). By historicism, Popper means the view of various influential social philosophies that historical prediction is the principal aim of the social sciences. It is a view that assumes this aim is attainable by discovering the 'rhythms' or the 'patterns', the 'laws' or the 'trends' that underlie the evolution of history (1959, p. 3)." It is the doctrine that "the task of science in general is to make predictions, and to put them upon a more secure basis. In particular, it is the task of the social sciences to furnish us with long-term historical prophecies."

Popper claims that "sweeping historical prophecies," like those provided by Marx, are "entirely beyond the scope of scientific method." In *The Open Society*, he tries to show that, despite their plausibility, they are based on a gross misunderstanding of the method of science. In particular, they neglect the distinction between *scientific prediction* and *historical prophecy*. Popper's principal aim is to show that "this prophetic wisdom is harmful. The metaphysics of history, he contends, impedes application of the piecemeal methods of science to the problems of social reform. "The future depends on ourselves." "[W]e do not depend on any historical necessity." . . . "We may become the makers of our fate only when we have ceased to pose as its prophets" (Popper 1966a, pp. 3-5).

4. Was Marx Actually a Historical Determinist?

Did Marx actually believe in such inexorable laws of history? And, what is at stake in this question?

The crux of Popper's criticism of historicism is that if we believe scientific laws predetermine our destiny, we are relieved of our responsibility for *making* the future better. We have no fundamental ethical choices to make, no agonizing policy options to think through, no agonizing weighing of advantages and disadvantages. If communism is in fact man's inevitable future, our only choice is whether or not to help ease the birth pangs.

There has been much debate about whether Marx actually was a historical determinist.[3] We will not venture too far into these debates, since they are inconclusive. Those who defend Marx against charges of historical determinism usually argue that if we take his work as a whole, or if we read him correctly, we will see that *he was not a historical determinist*. On the other hand those, who contend that Marx *was a*

---

[3] For a careful exploration of the distinctions between "laws of nature," "laws of history," and "historical laws" that addresses both Marx and Popper on these issues, see Berry (1999, pp. 121-137).

historical determinist have no difficulty finding strong textual support for such a view.

Perhaps the most important argument supporting the view that Marx was not an historicist is that he nowhere states any historical laws. Antony Flew, certainly no Marxist, points out that, after having stated that the aim of his work was to find historical laws, he fails to produce any (Flew forthcoming, p. 2). Nor does Popper, having claimed that Marx believed in inexorable laws of history, actually identify any such laws and hold them up for scrutiny.

According to John McMurtry, "Marx, himself, never speaks of "laws of history" (as opposed to laws of capitalism and its transition)" (1984, p. 31). Similarly, James Farr points out that what Marx calls "economic laws of history" are not laws *of* history, but are rather *"historical laws."* Farr gives as examples "the absolute law of the production of surplus value" and "the law of the tendency of the rate of profit to fall. That is, they are laws which express with social validity the conditions and relations of a definite historically determined mode of production," (1986, pp. 213-216) laws describing a historical process. Finally, Meiksins-Wood argues that the "proposition that history is simply the inexorable progress of productive forces is vacuous and by itself inconsistent with Marx's analysis of capitalism" (1984, p. 101).[4]

Furthermore, upon examination, Marx's sweeping statements of historical tendencies, despite their dramatic appearance, turn out to be so void of empirical content that they can hardly be called laws of history. Though Marx declares them to be inevitable, he says nothing substantive about the nature and dynamics of socialism and communism.[5] He defines them almost entirely in terms of *not being what capitalism is.* Nor does Marx provide any kind of concrete account of *how the revolutionary process will lead from capitalism to socialism, and then to communism.*

Nevertheless, those who argue that Marx *was* a historical determinist can find ample support for such a view in Marx's writings. For example,

---

[4] Rozov (1997, pp. 336-352) has also put forward thought-provoking criticism of what he understands to be Popper's views on theoretical history. Like Flew, McMurtry, Farr, and Meiksins-Wood, however, Rozov misses Popper's target. The historicism that alarmed Popper was the prophetic rhetoric within which Marx packaged his discussion of "natural laws of history . . . the more sweeping prophecies [which] are on a rather low intellectual level . . . compared with Marx's own high standards" (1966b, p. 197).

[5] As Meiksins-Wood writes: "The formula [about the contradiction between forces and relations of production] is fruitful as an account of capitalism, as a general law of history, it is rather empty and is not rendered more informative by the teleological proposition that capitalism emerged because history requires the development of productive forces and the development of productive forces requires capitalism" (1984, pp. 103-104).

Marx's prefaces to *Capital* contain strong evidence that he saw himself as a historical determinist. Here, in presenting his mature work, Marx states what he considers to be the main achievement of that work in no uncertain terms. He describes this achievement as the discovery of "tendencies working with iron necessity toward inevitable results." In Marx's words:

> Intrinsically, it is not a question of the higher or lower degree of development of the social antagonisms that result from the natural laws of capitalist production. It is a question of these laws themselves, of these tendencies working with iron necessity toward inevitable results. The country that is more developed industrially only shows to the less developed the image of its own future. (1967, pp. 8-9)

"Working with iron necessity toward inevitable results!" Could there be a stronger articulation of belief in historical determinism? Furthermore, Marx continues, even if we discover society's natural laws of motion, we can do nothing to change them. Again, in his own words,

> even when a society has got upon the right track for the discovery of the natural laws of its movement – and it is the ultimate aim of this work, to lay bare the economic law of motion of modern society – it can neither clear by bold leaps, or remove by legal enactments, the obstacles offered by the successive phases of its normal development. But it can shorten and lessen the birth pangs. (1967, p. 10)

In his Preface to the Second Edition of *Capital*, Marx quotes at length a review of the First Edition, published in the *European Messenger* (St. Petersburg). The passages *cited by Marx* describes his project in unmistakably historicist terms. "Of still greater importance to him [Marx]," writes the reviewer, "is the law of their variation, of their development, i.e., of their transition from one form into another, from one series of connections into a different one" . . .

> Consequently, *Marx only troubles himself about one thing, to show the necessity of successive determinate orders of social conditions*, and to establish, as impartially as possible, the facts that serve him for fundamental starting points. For this it is quite enough if he proves, at the same time, both the necessity of the present order of things, and the *necessity of another order into which the first must inevitably pass over;* and thus all the same, whether men believe it or do not believe it, whether they are conscious or unconscious of it. *Marx treats the social movement as a process of natural history, governed by laws not only independent of human will, consciousness and intelligence, but rather, on the contrary, determining that will, consciousness and intelligence.* (Marx 1967, p. 18, emphasis supplied)

... The scientific value of such an inquiry [into economic life] lies in the disclosing of the special laws that regulate the origin, existence, development, and death of a given social organism and its replacement by another and higher one. And it is this value, that, in point of fact, Marx's book has. (*European Messenger*, qouted by Marx in the preface of the second edition of *Capital*, 1967, pp. 18-19)

Marx does not take issue at all with the reviewer's synopsis. On the contrary, he quotes it approvingly, calling it a "generous way" of describing "the dialectic method." His only complaint is that the reviewer calls him an idealist in the German tradition. Marx then goes on to defend himself against this charge by showing how he thinks he has cleaned up the Hegelian dialectic, doing away with its mystifying side (1967, p. 19).

It is significant that Marx would choose to emphasize the historicist character (in Popper's sense) of his project several years after the initial publication of *Kapital*, in this Second Preface. It is strong evidence that this is how he understood his own work. How else are we to understand such statements? Anyone who wishes to argue that Marx did *not* believe in such laws of historical development should be able to explain why we find statements like these throughout his writings, as well as in what Engels wrote about his beliefs and intentions.

We cannot, thus, give a conclusive answer to the question of whether or not Marx was a historicist as Popper charges. It depends which aspects of Marx's writings one chooses to emphasize. Since Popper does not discuss such matters, or make his charge contingent upon them, the charge is wanting.

In the next section, we will examine more closely some flaws in Popper's critique of Marx's alleged historicism. I will lay the groundwork for the discussion of the last section. This last section will zero in on those aspects of Popper's critique which, I believe, are sound. These criticisms, I will conclude, bite hard on Marx's theoretical system.

5. Flaws in Popper's Critique

5.1. *Is Popper's Critique of Marx Itself Historicist?*

Popper holds Marxian doctrine responsible for such consequences as totalitarianism, unfruitful social science, the failure of Marxists to develop social technologies, a decline of faith in reason, and a fatalist attitude towards politics. Following Popper's anti-essentialist description of science, (e.g., in 1966b, pp. 9-26, 1974, pp. 12-23) scientific laws must always be taken together with *initial conditions* to obtain predictions.

Yet, in discussing Marx's doctrines and their influence, Popper seems to be ascribing historical causality to these ideas. He does not specify initial conditions which must be present for these doctrines to have the adverse effects he attributes to them. He also fails to acknowledge that there may be conditions under which these doctrines will not produce such effects. Moreover, Popper's argues that ideas do not control their own interpretation. It is those individuals who have adopted, interpreted, and applied Marx's ideas who are responsible for the views and actions they have based on them.

Marx's ideas have certainly not controlled all Marxist thinkers. On the contrary, many Marxists have subjected these ideas to both internal criticism and the test of experience, and modified them accordingly. Finally, political activists, movements, and parties inspired by Marx have interpreted his ideas in many different ways. Moreover, many applications of Marx's ideas have not led to the kinds of negative consequences Popper attributes to them.

In fact, individuals and movements inspired by Marx have often been in the forefront of the struggle for the open society. Popper recognizes this. For example, he writes of the "great movement" of the workers of Vienna "led by the social democratic party." Even though he regards the "Marxist historicism of their social democratic leaders as fatally mistaken," he goes on to praise these leaders for being "able to inspire them with a marvelous faith in their mission, which was nothing less, they believed, than the liberation of mankind . . . [The] whole movement was inspired by what can only be described as an ardent religious and humanitarian faith . . . It was an admirable programme" (1974a, p. 27). Popper also writes that, during the interwar period the social democrats, where they still existed on the continent, "were after all, the only political force still resisting tyranny" (1974a, p. 90). In short, Popper not only recognizes that such a positive political program can result, not only despite the social democrats' Marxism, but *because of* it.

Moreover, fatalism is hardly an appropriate characterization of the actual behavior of parties influenced by Marxism. Socialist parties have, since the late 19th century been in the forefront of those fighting actively for the what Popper called "the open society." If anything, the charge of fatalism would appear to better fit the attitudes and behavior of some liberal and conservative parties.

Finally, as E.H. Carr has pointed out, Popper's assertion that "'in human affairs everything is possible' is either empty or false"

(1964, p. 93). In human affairs, there are limits to change. It may be not *fatalistic*, but rather *realistic* and *responsible* not to try to change what cannot be changed. Popper knows this. In fact, one of his criticisms of Marx is that he encourages a prejudiced view that excludes hypotheses stating limits to human action. The real problem in decision making about human affairs is to know the difference between what can and cannot be changed.

5.2. *What Is the Responsibility (If Any) of Marxism for the Downfall of Democracy and Rise of Totalitarianism?*

Marxist historicism may in fact help to account for some of the tragic mistakes of political parties in the Marxist tradition. Moreover, the horrors of Stalinist totalitarianism did come wrapped in Marxist-historicist jargon. However, Popper does not spell out causal linkages between Marxism and the negative consequences he attributes to it. Moreover, on closer examination, it is not clear to what extent such linkages actually exist. After Marx's death, with the passage of time, European workers increasingly organized themselves into socialist parties and trade unions. These were very successful in gradually bettering the lot of their members. Partly in consequence of this, the mainstream of Marxian socialism turned increasingly away from revolutionary action toward pragmatism, reformism, and moderation.

Part of the Marxian socialist movement became revisionist. Another part, though officially retaining its revolutionary credo and rhetoric, became evolutionary and reformist in practice. Even in Russia, the mainstream of Marxian socialism was not revolutionary in its politics. Taking Marx more seriously than would Lenin, they believed Russia had first to pass through a phase of capitalist development before a socialist revolution could take place. For them, the task of Marxists was to support liberal parties in their struggle for the establishment of a constitutional monarchy and a capitalist economy.

World War I was a devastating shock for the moderate mainstream of the socialist movement. Contrary to their deeply-held ideological expectations, socialist parties throughout Europe supported the war efforts of their national governments. Many socialists believed that it was the abandonment of the revolutionary principles of socialism that had been responsible for the failure of socialist parties to prevent the War. At the same time, it looked like the dramatic victory of the Bolsheviks in Russia had been due to their uncompromising revolutionism. This led to a loss of confidence in mainstream socialism, and to radicalization of part of the socialist movement (see, e.g., Borkenau 1962, pp. 57-107).

Without these extraordinary initial conditions, the historicism in the socialist tradition may well have remained harmless, if not actually beneficial.

Popper, in his autobiography, refers to violence instigated by *communists* to illustrate the alleged penchant of Marxism for violence. He writes of the "dogmatic character of the [Marxist] creed, and its incredible intellectual arrogance" (1974, pp. 25-26). Here, he misses the fundamental distinction between communists and social democrats. He also ignores the impact of the desperate crisis situation in Europe in the interwar period, not only on Marxism, but on political belief and action in general. Moreover, by Popper's own account, the communists were not Marxian determinists at all. Following Lenin, they had replaced Marx's determinist view of history with a voluntaristic Vulgar Marxism. Rather than the Revolution being brought about by historical forces, it was the will of an elitist ruling party that determined the course of history. Furthermore, Bolshevism and Soviet Communism cannot be understood without taking into account such factors as the deep entrenchment of despotism in the Russian political tradition[6] and Lenin's fusion of the Russian revolutionary tradition with Marxist ideas (see, e.g., Fainsod 1967, pp. 2-86; Borkenau 1962, pp. 22-56).[7] Before the Russian Revolution, Lenin's Bolsheviks were a marginal political sect. They were able to come to power only due to the political vacuum left by the collapse of the the tsarist regime. The tsarist autocracy had managed to

---

[6] As Szamuely writes: "It was in the reign of Ivan IV that the previously blurred outline of a State commanding the totality of powers, with unlimited and unquestioned authority over every subject, took on clear and definite shape . . . Comparing the absolute monarchy in Russia with those of Western Europe," he continues, "In Russia, the only rights that remained were those of the State, and these were total and limitless (1974, p. 35).

[7] Moreover, differences among those who claim to be the heirs of Marx's legacy have often been more fundamental than differences dividing some Marxists from some non- and anti-marxists. Who, for example, should be regarded a more faithful follower of Marx, the "revisionist," Edouard Bernstein, or the "orthodox," Karl Kautsky? Bernstein, like Marx wanted theory to accord with practice, and tried to save Marxism by bringing it up to date, or the "orthodox" Marxist, Karl Kautsky who, although widely recognized as the most important protagonist of Marxist doctrine after Engels' death, emasculated Marxist theory, effectively reducing it to a verbal radicalism. And what about Lenin and his successors? If Lenin's followers were not so numerous, if they did not wield so much political influence and the means of making their interpretations of Marx so loudly and widely heard, how seriously would their claim to Marx's intellectual mantel be taken? Could it not be argued that Lenin and, even more, Stalin went far beyond Berstein's revisions and Kautsky's emasculation to the final destruction of theoretical Marxism? I certainly do not think that the best Marxist is the most stubborn and dogmatic apologist for any particular theories and methods articulated in Marx's works.

retard significantly the natural development of political parties and movements in Russia. Only the Bolsheviks, with their fanaticism, rigid discipline, underground organization and conspiratorial methods were strong enough to take power.

One might, argue that not much of Marx remains in Lenin's and Stalin's renditions of his theories. Abdurakhman Avtorkhanov, for example, contends that Lenin "buried without a trace the famous 'laws' of Marxist dialectics."

> Against Marx's "immanent" laws of the natural-historical process, Lenin set the conscious, organized will of revolutionary creativity, the "will to power." A Nietzschean to the core, Lenin, however, never let himself seem to be an opponent of Marx; he did everything in the name of Marxism. (Popper 1966, p. 6)

Finally, one cannot explain the concrete incarnations of totalitarianism that are the target of Popper's attack without taking into account other important factors. Among these are, for example, the exceedingly cruel and paranoid personalities of Stalin and Hitler and the desperate situations that helped bring them to power. Moreover, we must also take into account the 20th century technologies that multiplied the horrors of these regimes.

5.3. *Cannot "Pernicious Doctrines" Be Tamed? Might They Not Be Necessary?*

The strains of thought Popper sees as leading to totalitarianism (historicism, romanticism, collectivism-holism) have existed in benign forms in many regimes. Horrifying consequences may indeed ensue when some variants of these strains of thought become the supreme values and dominating visions of ideologies. This is particularly true when they are joined to dogmatic, intolerant attitudes and monopolistic control of state power. However, not all incarnations of such strains of thought need be ugly and virulent. Some variants of Marxism are balanced and tempered by other ideas in actually existing belief systems. Popper ignores this. He also fails to discuss the possibility that such ideas may contribute positively to the creation and maintenance of community among individuals. Finally, he does not consider that the danger of such ideas may depend on other factors in the political environment.

For example, Popper asserts that "all nationalism or racialism is evil" (1974, p. 83). He appears to treat the two as synonymous. Yet some varieties of nationalism have been neither evil nor harmful. Concrete political cultures consist of mixtures of ideas and attitudes. The national idea may coexist and blend in with other ideas, like respect for the rights of individuals and of other nationalities. Moreover, some forms of nationalism are not racist, but are based on culture or on the state. Some varieties of nationalism, sometimes even racialist-based nationalism, have been neither militant nor exclusive.

To be sure, there have been pernicious and disastrous interpretations of nation and of other holistic ideas, like class. It does not follow, however, that such interpretations are the only ones possible. Popper does not demonstrate that there cannot exist nationalist doctrines or programs in which the value of the nation is tempered and balanced by other principles. He pays no attention to concrete political traditions in which such doctrines have been present in harmless, even beneficial, forms.

We can make a similar point about the role of historicist imagery in politics. Popper's warnings about the dangers of historicism are powerful. He, nevertheless, fails to show why such ideas cannot be tempered and balanced by other ideas in actual political practice. President John F. Kennedy's Inaugural Address provides an example of how reference to the "court of history" can be tempered by simultaneous emphasis of the importance of individual conscience and responsibility.

> With a good conscience our only sure reward, with history the final judge of our deeds, let us go forth to lead the land we love, asking His blessing and His help, but knowing that here on earth, God's work must truly be our own. (Sorensen 1966, p. 278)

On the other hand, cruel and destructive tyrannies have existed that have not been based on the strains of thought Popper sees at the heart of totalitarianism.

5.4. *Was Marx a Moral Positivist?*

By moral positivism, Popper means the view that no moral standard is possible but the one which exists. In other words, for example, might makes right. Yet Popper asserts: "I feel sure that had he considered these implications, Marx would have repudiated historicist moral theory." "Numerous remarks and numerous actions prove," Popper writes, "that it was not a scientific judgement but a moral impulse, the wish to help the oppressed, the wish to free the shamelessly exploited and miserable workers, which led him to socialism" (1966b, p. 207).

Something seems wrong with this argument. Popper, himself, provides ample argument to the effect that Marx saw the future he predicted as morally superior to the system he described as doomed. It follows that neither Marx's moral condemnation of capitalism, nor his moral approval of the inevitable future to come, required any specific ethical argumentation. They were self-evident to Marx, as the principles of "life, liberty, and the pursuit of happiness" were for the signers of the American Declaration of Independence. As Robert Tucker points out, there is no evidence that "the principle of historical inevitability plays the part of a basic ethical norm in Marx's scheme of thought." In other words,

> There is nothing to indicate that Marx morally affirmed the future world revolution on the ground of its presumed inevitability. Far from deciding that a communist revolution would be desirable after discovering that it was inevitable, he became convinced as a young man of its desirability and then embarked on a life-long effort, materialized in *Capital*, to prove that it must come. (Tucker 1961, p. 21)

Moreover, Tucker argues that it would be difficult to explain the "burning intensity of Marx's moral rejection of the existing world" merely by "the intellectual perception or proof that it is doomed to give way to a different order in the future" (1961, p. 21). The moral value of the classless society and the "liberation of man from the brutalizing forces that had previously thwarted the free and full development of their potentialities" (Gardiner 1967, p. 521) were self-evident to Marx. The fact that he was mistaken about the future is quite a different matter.

5.5. *Who Is Not a Historicist and a Victim of Scientism? Why Single out Marx?*

"To appreciate the Marxian conception of history," writes George Lichtheim, "it is necessary to remember that its author belonged to an age which recognised no limitations to the range of knowledge available to a single mind. Historical generalizations of the most far-reaching and universal kind were not merely admired but expected" (Lichtheim 1965, p. 141). We can make similar observations about Marx's scientism. Popper attributes to Marx a flawed understanding of the nature of science. However, by Popper's own standards, faulty views of science underlie most of modern social science, as well as most of modern liberal political thought.

By Popper's own definitions and standards, most modern social and political thinkers are historicists and victims of scientistic thinking. Popper criticizes several of them specifically in *The Open Society*. Some he criticizes even more severely than Marx, for their historicist ideas – Mill, Comte, Weber, Durkheim, Mannheim, among others. For example,

Popper asserts that Marx, in claiming "'to lay bare the laws of motion of modern society' might be said to carry out Mill's programme: 'The fundamental problem . . . of the social science, is to find the law according to which any state of society produces the state which succeeds it and takes its place'" (1966b, p. 87).

Although Hobbes's scientism may be individualistic, it is not particularly supportive of humanitarianism and individual freedom. To be sure, it was moderated by Locke. However, even in Locke, there remains a scientistic base of Newtonian mechanism that can support a hard, mechanistic liberalism. Bentham's pleasure-pain calculus provides a scientistically-based notion of justice. To be sure, it is individualistic. However, it can justify the pain of some individuals if offset by the pleasure of the majority. We might also point to B.F. Skinner's *Walden Two*, which, although individualistic and benevolent, appears as a kind of scientistic totalitarianism. To be sure, Mill and others modified Bentham's views, but the scientistic base still remains. It is possible to identify in the theoretical views of most modern political thinkers who take their bearings from their beliefs about modern natural science.

So, a closer examination of Popper's charges against Marx shows them to be unclear and less than conclusive. Popper's definition of historicism is so vague that it is difficult to determine which thinkers fall into this category and which do not. He does not explain how so much of liberal thought can be drenched with historicist and scientistic ideas without sliding down the slippery slope into totalitarianism. More generally, he fails to specify conditions under which historicist and scientistic ideas become dangerous.

6. Where Popper's Critique Bites

I now propose to zero in on what I think are Popper's most important criticisms of Marx's. These are his criticism of its obscurantism and hardy resistance to correction by experience.

Marx's definition of communism, in *The German Ideology*, provides good illustration. "Communism," Marx writes, "is for us not a state of *affairs* which is to be established, an *ideal* to which reality [will] have to adjust itself. We call communism the *real* movement which abolishes the present state of things. The conditions of this movement result from the premises now in existence" (Marx and Engels 1970, p. 56-57).

What does it mean to "abolish the present state of things"? Obviously, any given state of things consists of some elements that existed previously and some new elements. Some elements of any state of things will persist, while others will disappear. Sometimes, when important novelties appear, or when important characteristics of a state of things disappear, we may want to speak of "a new state of things." Or, we might wish to speak of a new *stage* in the development of this state of things.

The expression "abolish the present state of things" obviously cannot mean abolition of all of its features. This would mean that nothing at all would remain of society. Marx obviously does not mean this. He means abolition of what he has defined as essential features of "the present state of things," that is, capitalist productive relationships. However, this leaves the statement almost vacuous. Marx defines communism as whatever "real movement" (an expression that has little if any empirical content) "abolishes the present state of affairs" (whatever this may mean).

Or, let us take another example: Marx insisted that it is *men* who "produce principles, ideas and categories." This statement is vague and ambiguous. It says nothing about *how* men produce these principle, ideas and categories. To be significant, it would have mean that men have some degree of freedom in producing and failing to produce ideas. However, Marx also wrote that it "is not the consciousness of man that determines his existence. Rather, it is his social existence that determines his consciousness" (Marx 1904, p. 11-12). If it is man's social existence that determines his consciousness, in what sense can we truly regard him as a producer of principles, ideas and categories? Why would he not be just a mere instrument in the productive process?

Marx's theory of alienation represents private property as the result of alienation. Moreover, in Marx's description of the alienation of labor under capitalism, this certainly appears to be the case.

> [I]f the product of his labor, his labor *objectified*, is for him an *alien*, hostile, powerful object independent of him then his position toward it is such that someone else is master of this object, someone who is alien, hostile, powerful, and independent of him.
>
> . . . Just as [man] creates his own production as the loss of his reality, as his punishment; his own product as a loss, as a product not belonging to him; so he creates the domination of the person who does not prooduce over production and over the product. Just as he estranges his own activity from himself, so he confers to the stranger an activity which is not his own.
>
> . . . *Private property* is thus the product, the result, the necessary consequence of *alienated labor*, of the external relation of the worker to nature and to himself. (Marx 1964, pp. 116-117)

So, as long as capitalist social relations survive, man remains alienated. Conversely, once capitalist social relations have been abolished, man will no longer be alienated. This state of affairs is, by definition, communism.

> *Communism* [in its final form will be] the *positive* transcendence of *private property*, as *human self-estrangement*, and therefore as the real *appropriation of the human* essence by and for man; communism therefore as the complete return of man to himself as a *social* (i.e., human) being ... This communism, as fully-developed naturalism, equals humanism, and as fully-developed humanism equals naturalism; it is the *genuine* resolution of the conflict between man and nature and between man and man – the true resolution of the strife between existence and essence, between objectification and self-confirmation, between freedom and necessity, between the individual and the species. Communism is the riddle of history solved, and it knows itself to be this solution. (Marx 1964, p. 135)

All this has a plausible ring. Marx defines communism as the opposite of the concrete situation of labor relations in the unrestrained capitalism of the 19th century. The undeniable reality of this concrete situation gives rise to the illusion that its opposite, communism, must also be real. However, this illusion dissolves upon reflection.[8] Marx says nothing concrete about communist social or economic relations. He presents no picture at all of communism as an economic system, the functioning of which we might plausibly envision. Socialism and communism turn out to be illusions created by Marx's dialectics.

Marx describes alienation as resulting from the essence of the system – that is from capitalist social relations. All other hypotheses regarding the nature and causes of alienation are, thus, ruled out *a priori*, by definition. "Only the evolution of the underlying essence, the economic reality, can produce any essential or real change – *a social revolution*" (Popper 1966b, p. 109).

Marx imposes solutions to all kinds of important empirical questions by definitional fiat. What is it that causes alienation? Can we really just

---

[8]As Dunn writes: "Marx's own picture of the political promise of revolution at least in advanced capitalist societies remained premised with some firmness on the collapse of the French Ancien Regime in 1789, a structure of social arrangements with its own long-cherished ideology, collapsing beneath the weight of its own sheer ideological implausibility ... But since it was such a blatant characteristic of what did actually happen in 1789, ... it was as natural for Marx to extrapolate it from those events as it was for example for Paine to have done so at roughly the time that they occurred" (1979, p. 87).

assume that capitalist social relations are the cause? Doesn't this definition of the problem of alienation assume what Marx would have to demonstrate? The language obscures the fact that crucial argumentative links are missing. It immunizes the theory against attempts to point out such flaws.

This is the heart of Popper's critique of Marx. It has serious consequences for those interested in defending and improving Marxism.

What if we accept the view that Marx was, in fact, interested in changing society? What if he truly did *not* believe that developmental laws govern the course of history? Would he not have been obliged to say something about what might be done to bring about a better and more just society? Yet, Marx scorns all attempts to improve society by piecemeal reforms. The only action he recommends is revolutionary action. This is because, in his view, the only real cure for ills he identifies is the Revolution that brings about a change in the essence of the system.

Marx provides a scathing diagnosis of the injustices of Capitalist society. He shows, not only that capitalism is doomed, but why it deserves to be doomed. Yet, the only remedy he has to offer is revolution. Communism, which he vaguely represents as the opposite of doomed capitalism follows from this. Social technology is not only absent, but there are arguments for avoiding it.

7. Conclusion

Popper is right in characterizing Marx's writings as a mixture of science and religion. To be sure, the place of Marx's works among the classics of social science is unquestionable and well-deserved. However, without their religious-historicist dimension, Marx's ideas would never have become the ideological backbone of one of the most influential political movements in history. A theory does not capture people's hearts and minds just by being good science. It has to offer something worth believing in. Yet few will believe in any wild promise of a better life. People must also be convinced of the veracity of the promise. In our times, it is science that is the touchstone of truth. If something is believed to be backed by the alleged authority of science many people assume it must be true.

The socialist vision is much older than Marxian socialism. Moreover, there have been other socialisms that failed to achieve anywhere near the political success of Marxism. Marxism combines its promise of a better future with the alleged warrant of science that this future is assured. It is this combination of qualities that has given Marxism the ability to win the allegiance of so many people over the past century and a half.

Popper penetratingly exposes the pseudo-scientific foundations of Marx's prophesy. His critique of Marx is a treasure chest of insights into the social scientific enterprise, the value of which extends far beyond the exposure of Marx's errors. Popper also draws attention to consequences that can follow from dogmatic faith in such false prophesy.

Yet, when we attempt to pin down causal linkages between Marx's writings and Stalinist totalitarianism the target proves to be elusive. Popper's charges against Marx are poorly focused. This makes it easy for Marx's sympathizers to dismiss Popper's critique without grasping its import.

Popper may be right that Marxist historicism can seduce weak minds. However, weak minds may also believe in Marxist historicism without any negative consequences at all. On the other hand, frustrated, angry people may perpetrate or participate in reigns of cruelty and terror without any assistance from Marxism or, for that matter, any historicist theory. Weak minds, especially in desperate situations, can latch onto ideologies much cruder than Marxism.

At the same time, prophesy, whether methodologically sound or not, need not necessarily be dangerous. Indeed, it often serves as valuable social criticism. Finally if, as by Popper's own testimony, historicism and scientism are rampant even within the tradition of liberal political thought, it should lead us to wonder all the more about the nature of the link between historicism and totalitarianism (Eidlin 1997, p. 15-16).

What then completes the causal link between Marxism and totalitarianism? I think it is an attitude of dogmatism and intolerance. It cannot be denied that dogmatic believers in the marxian doctrines that Popper critically identifies, have all too often turned a deaf ear to the suffering of victims of their revolutionary violence. It cannot be denied that belief in Marx's promise of an inevitable communist future has all too often led to a willingness to sacrifice recklessly the delicate structures that sustain freedom and civility.

Yet is it right to blame Marx and his writings for consequences due to dogmatic and intolerant attitudes? Is it right to hold an idea or its author responsible for the interpretations and actions of those who believe in it? As we have noted, most marxists have not been totalitarians, not even most of those who espoused totalitarian versions of marxist ideology. In fact, marxist ideals along with purist marxist opposition appear to have been among the factors that led to the softening and downfall of Communist regimes.

On the other hand, as we can observe, for example, in the movement for political correctness, even such high-minded humanitarian principles as anti-racism, and opposition to sexual and gender harassment, when joined to dogmatism and intolerance, take on a totalitarian flavor. Even the philosophy of Sir Karl Popper, despite the high value it places on tolerance, openness to criticism, and respect for other people, can take on a totalitarian flavor when combined with an attitude of dogmatism and intolerance. Tolerance and openness to criticism as well as dogmatism and closed-mindedness are attitudes more than they are doctrines. As such, they are more closely bound up with a person's character and psychology than with the ideas that person publicly espouses.

University of Guelph
Department of Political Science
533 Mackinnon Building
Guelph, Ontario NIG 2W1
Canada
e-mail: feidlin@uoguelph.ca

REFERENCES

Avtorkhanov, A. (1966). *The Communist Party Apparatus*. Cleveland, OH: World Publishing Co.

Berlin, I. (1963). *Karl Marx*. Third Edition. London, New York: Oxford University Press.

Berry, S. (1999). On the Problem of Laws in Nature and History: A Comparison. *History and Theory* **38** (4), 121-137.

Borkenau, F. (1962). *World Communism*. Ann Arbor, MI: The University of Michigan Press.

Carr, E.H. (1964). *What is History?* Harmondsworth: Penguin.

Cornforth, M. (1968). *The Open Philosophy and the Open Society: A Reply to Dr. Karl Popper's Refutations of Marxism*. New York: International Publishers.

Dunn, J. (1979). *Western Political Theory in the Face of the Future*. Cambridge: Cambridge University Press.

Eidlin, F. (1983). Isn't Popper the Best Marxist? *Newsletter for Those Interested in the Philosophy of Karl Popper* **1** (3/4), 18-20.

Eidlin, F. (1997). Blindspot of a Liberal: Popper and the Problem of Community. *Philosophy of the Social Sciences* **27** (1), 5-23.

Fainsod, M. (1967). *How Russia is Ruled*. Revised Edition. Cambridge, MA: Harvard University Press.

Farr, J. (1986). Marx's Laws. *Political Studies* **34**, 202-222.

Flew, A. (unpublished). Prophecy or Philosophy. Historicism or History? Unpublished manuscript.

Gardiner, P. (1967). Speculative Systems of History. In: *The Encyclopedia of Philosophy*, vol. 7. New York: Macmillan.

Jarvie, I.C. (1982). Popper on the Difference between the Natural and the Social Sciences. In: Paul Levinson (ed.), *In Pursuit of Truth*, pp. 83-107. New York: Humanities Press.

Lichtheim, G. (1965). *Marxism: An Historical and Critical Study*. Second Edition (revised), New York: Praeger.

McMurtry, J. (1984). Reply to deLaunay and Eidlin. *Newsletter for Those Interested in the Philosophy of Karl Popper* **2** (1/2).

Marx, K. (1904). *A Contribution to the Critique of Political Economy*. Chicago: Charles H. Kerr & Company.

Marx, K. (1964). Economic and Philosophical Manuscripts of 1844. In: *Early Writings*, pp. 61-219. New York: International Publishers.

Marx, K. (1967). *Capital: A Critique of Political Economy*, vol. I. New York: International Publishers.

Marx, K. and F. Engels (1970). *The German Ideology*. New York: International Publishers.

Meiksins-Wood, E. (1984). Marxism and Historical Progress. *New Left Review* **147**, 95-108.

Popper, K.R. (1961). *The Poverty of Historicism*. Second Edition, London: Routledge and Kegan Paul.

Popper, K.R. (1966a). *The Open Society and Its Enemies I: The Spell of Plato*. Princeton, NJ: Princeton University Press.

Popper, K.R. (1966b). *The Open Society and Its Enemies II: The High Tide of Prophesy*. Princeton, NJ: Princeton University Press.

Popper, K.R. (1974). Intellectual Autobiography. In: P.A. Schilpp (ed.), *The Philosophy of Karl Popper*, pp.3-181. La Salle, IL: Open Court.

Rozov, N.S. (1997). An Apologia for Theoretical History: In Memory of Sir Karl Raimund Popper. *History and Theory* **36** (3), 336-352.

Sorensen, T.C. (1966). *Kennedy*. New York: Bantam Books.

Suchting, W.A. (1985). Popper's Critique of Marx's Method. In: Gregory Currie and Alan Musgrave (eds.), *Popper and the Human Sciences*, pp. 147-163. Dordrecht: Martinus Nijhoff.

Szamuely, T. (1974). *The Russian Tradition*. London: Morrison and Gibb.

Tucker, R. (1961). *Philosophy and Myth in Karl Marx*. Cambridge: Cambridge University Press.

Bryan Magee

POPPER'S PHILOSOPHY AND PRACTICAL POLITICS

Some years acquire symbolic status, and one such year is 1968. All over Europe and the United States university students exploded into violent rebellion. Insofar as this would-be revolution had an ideology it was unquestionably Marx-inspired, even if the marxism was not always orthodox. It so happens that in the years 1970-71 I was teaching philosophy at Balliol College, Oxford, and because of Oxford University's system, almost unique, of individual tuition for undergraduates, this meant I found myself in a continuing one-to-one relationship with bright students who were in the throes of revolutionary fervor.

Arguing with them was enormously illuminating for me. It seemed as if the more intelligent they were the more passionately marxist they were – but also the more affected they were by intellectually serious criticisms of marxism, which usually they were hearing for the first time. It was when they found themselves unable to meet these that they revealed where their fundamental motivation lay. This was not usually a positive one of belief in marxist ideas. Still less was it commitment to communist forms of society, which usually they had been defending without knowing anything about the reality of them. The motivation was usually negative: it was inability or refusal to come to terms with their own society as they saw it. Psychologically, this was nearly always at the root of their attitude.

Basically the chain of cause and effect between their ideas seemed to go something like this. They longed to live in a perfect society. But only too obviously the society in which they found themselves contained serious evils. So this form of society had to be rejected. A particularly interesting point here is the fact that, because what they demanded was perfection, they thought that if anything was seriously wrong then the whole must be rejected. If, say, newspapers reported cases of old and

In: E. Suárez-Iñiguez (ed.), *The Power of Argumentation* (*Poznań Studies in the Philosophy of the Sciences and the Humanities*, vol. 93), pp. 55-69. Amsterdam/New York, NY: Rodopi, 2007.

poor people dying of hypothermia in winter because they had no heating in their homes, the students would say savagely "there's something sick about a society that lets old people freeze to death in the winter." If there were reports of students unable to take up university places because of an inability to get grants they would say "there's something fundamentally rotten about a society that refuses to educate people unless they've got money." It was virtually a formulaic response, of the fixed form "There's something fundamentally rotten about any society in which x happens," with x standing for any serious social evil. If anything at all was seriously wrong, the whole of society was sick: unless everything's perfect everything's rotten. Such an attitude could rest only on utopian assumptions. And it quite naturally made those who held it receptive to a holistic as well as systematic social critique of the only society they knew. It also led most of them to suppose, erroneously, that there must be something somewhere that was infinitely better: since, plainly, things were not perfect here, they must be perfect somewhere else – or, at least, people somewhere else must be trying. Criticisms of communist reality were nearly always met by the counter-accusation that things were just as bad here, if not worse, and at least the communists were striving to realize a moral ideal, which our cynical and self-interested politicians were not.

These attitudes display several errors of a fundamental character to which intelligent people in general are prone when they think about politics. Instead of starting from what actually exists, and trying to think how to improve it, they start from an ideal of the perfect society, a sort of blueprint in the mind, and then start thinking of how to change society to fit the blueprint. If they cannot see any practicable way of getting from reality to the blueprint they may be tempted then to think in terms of sweeping reality away, in order to start from scratch, in order to realize the blueprint.

Karl Popper's ideas are a marvellous antidote to such illusions. First of all he is insistent on its being an inescapable fact that wherever you want to go you have to start from where you are. Even the most cataclysmic revolution is an attempt to achieve certain ends, a way of trying to change society as it actually is into a different form of society that is preferred. And as the history of revolutions illustrates, existing society never is swept completely away: huge and important features of it always persist into the successor society, usually to the bafflement and chagrin of the revolutionaries. As a way of achieving desired social change revolution is exceedingly cost ineffective as well as ineffectual. First and foremost, large numbers of people get killed, or are made to suffer appallingly in other ways. Second, desirable as well as undesirable social fabric is destroyed. Third, unrestrained violence on a large scale is uncontrollable when accompanied by a breakdown in the social order. Fourth, because it is uncontrollable, the kind of society that emerges from it is nearly always one which the revolutionaries themselves say is quite different from what they wanted.

All forms of political thinking that start from blueprints of what is desired are anathema to Popper, and rightly so. All modern forms of society are in a state of perpetual change, and as time goes by the pace of this change gets faster, not slower. If we were to set ourselves the task of actualizing the most ideal blueprint, and then succeeded in actualizing it, even then change would not just suddenly stop. Marx and Engels thought it would – thought that with the realization of their perfect society history would come to an end. But nobody now believes this. Change will go on. So from the very moment we actualize our blueprint reality will start moving away from it and turning into something else. So the real political task is not to actualize an ideal state of affairs that can then be preserved for ever. This is the task to which the greatest political thinkers of the past, such as Plato and Marx, addressed themselves, but in reality it is not even an option. The real political task is to manage change.

As part of the process of perpetual change, people's aspirations and priorities perpetually change. So again, there too, even if we were able to start out with an ideal blueprint, and to succeed in our approach to it, as we worked towards it people's wishes would start moving away from it, so that even before we achieved it scarcely anybody would wholeheartedly want it. Something close to this has only too obviously happened in the late twentieth century with the ideal of socialism under its classic definition of public ownership and centralized planning of the means of production, distribution and exchange, an ideal which earlier in the century powerfully motivated millions of intelligent and well-meaning people, yet to which now scarcely anyone subscribes.

There is a need for perpetual revision of aspirations and goals, and this is inimical to the whole idea of a blueprint. Blueprints are fixed, static: if they changed unceasingly they would not be blueprints. They are therefore at best a source of never-ending problems, given the reality of permanent social change, and only too often they are a source of tragedy. Because they are fixed, people's attitudes towards them become fixed: they become objects of quasi-religious commitment and belief; and because they are seen as ideally desirable, political opponents who actively try to prevent them from being realized come to be looked on as wicked people who must be stopped, perhaps even removed from the scene altogether; and their elimination is seen as fully justified, indeed demanded, morally. Blueprints thus lead to rigidity, fanaticism, and through them to anti-rationality in many forms. The man with a blueprint usually knows he is right; and because of his utter certitude he feels justified in eliminating opposition by whatever means may be found necessary.

Popper's recommendation is that what we should eliminate are blueprints, eliminate them from our thinking entirely. Instead of basing our approach on an imaginary state of affairs that does not actually exist and is never going to exist, he recommends that we start from the social reality in which we find ourselves, and that we examine it critically to discover what is wrong with it, and to see how it may be improved. From that starting point he proposes what might be called a methodology for the management of change. I would like to go through this proposed method step by step.

First of all we are required to formulate our problems with care. That means, among other things, not taking for granted what they are. We have to ask ourselves what precisely are, say, the main problems that face us in the field of primary education? What, precisely, are the main problems that face us with the treatment of teen-age offenders against the law? What, precisely, are the main problems that face us in our relations with the United States? And so on and so forth.

There will, legitimately, be differences of opinion about what the problems are, before one has even begun to think in terms of solutions and these differences should be thoroughly debated. It is of the utmost importance to get diagnosis right before one proceeds to cure, otherwise the proposed cure will be the wrong one, not effective, quite possibly harmful. So a lot of time and trouble and thought and work needs to go into the identification and formulation of problems before one attempts to move forward from that position.

Once a problem has been identified and clearly formulated, the next step is to consider alternative possible solutions. At this stage especially there can be opportunity for great boldness, and also for imagination and ingenuity, for freshness of perception and vision, for unexpected initiative. Usually it is here, if anywhere, that creative politics comes in.

But of course many if not most of the proposed solutions would not actually, if tried, work out very well in practice. As soon as you start to do something, anything, unexpected snags arise. Even in the most apparently sensible undertakings measures take longer than expected, or cost more, or prove to be administratively cumbersome, or alienate some of the individuals involved, or have unfortunate side-effects.

It is of the utmost importance that these drawbacks should be minimized in practice by being foreseen and avoided. So proposed solutions need to be critically examined and debated, with the explicit object of bringing their faults to light before they are turned into reality. The more effective the criticism at this stage, the greater the saving in time, in economic resources and human happiness. The proposals whose effective criticism is most desirable, because most fruitful, are those of government because these are the ones that are put into practice on the largest scale, and with the most powerful backing, and with the greatest effect on peoples' lives. Full and free critical public discussion of proposed government policies is therefore essential if avoidable large-scale error is indeed to be avoided – without such discussion there will inevitably be more, and more costly, public-policy disasters than there need to be.

And of course there will be mistakes anyway. Even after a great deal of misplaced expectation has been eliminated by critical discussion, and the proposals thus critically improved are put into practice, things will still go wrong. Our actions have unforeseen consequences. So there is a need for practical as well as theoretical vigilance. After a policy has survived critical discussion and been put into practice, a critical eye needs to be kept on how it is actually working out, with a view to catching the first sign that it is not working as hoped. At this stage the most important thing is not to be seeking reassurance that all is well, but the opposite, to be on the alert for the possibility that things are not going as they should. This requires the practical monitoring of public policy in action, and for that to be effective people need to be free to criticize not only a government's proposals but also its deeds. Again, the sooner harmful practices are identified, the greater the saving will be in time, in resources and in human happiness. Governments that forbid public debate and criticism of their activities are bound to persist in mistaken, costly, and harmful practices for much longer than they otherwise would; and being government activities these mistakes will usually be on a large scale.

This, in its barest outline, is the methodology recommended by Popper to the practical politician. Some people may say it is embarrassingly obvious. I only wish it were. You do not need to be a very attentive reader of the serious press to realize that this is not how real-life politics is for the most part conducted. And, as someone who was a professional politician for nearly 10 years, I can assure you that the thought processes involved do not come at all easily to many politicians; indeed, some have serious difficulty in understanding them even when they are explained. If Popper's principles seem obvious to a philosophy-oriented audience it is because they are so rational, so congruent with situational logic. That is a powerful recommendation for them, but alas, it has not yet brought about their general acceptance or even comprehension. The task of actively promoting them still requires adherents.

Other critics may object that the whole approach is too cautious and therefore too slow. We haven't got time for all that talk, they may say: it's a luxury we can't afford. To this I believe the best reply is that of all possible political methods this is the one most likely to maximize the extent to which change remains under rational control. Attempts to short circuit processes of criticism are almost bound to lead to more error, and therefore more cost, and also more in the way of unintended consequences. There may indeed be more *change*, but disconcertingly much of it, too much, will be not in the required direction. This turned out to be one of the systematic shortcomings of centralized planning, and led in practice to its becoming almost invariably associated with systematized lying.

The approach advocated by Popper is a broad recipe for effective and successful problem-solving. As such it has a general application to most practical affairs, not only to politics but to administration in any form, and also to business. People familiar with his philosophy of science and his more general theory of knowledge will have noted already that it instantiates his formula for problem-solving in those fields

$$P_1 \longrightarrow TS \longrightarrow EE \longrightarrow P_2$$

where P_1 is the initial problem, TS the trial solution proposed to this problem, EE the process of error elimination applied to the trial solution, and P_2 the new situation thus arrived at, with its new and sometimes unexpected problems. In fact the relationship between Popper's methodology of politics and his theory of knowledge is so close that it is worth our going on now to look at some specific features that they have in common.

First, Popper regards himself in both fields as addressing not a static or stable state of affairs but a process of change, and he sees the main challenge as being how to manage change, in one case the growth of knowledge, in the other ongoing social development. In both cases he sees the demands this makes on us as consisting above all else of problem-solving. In both cases, therefore, he thinks we should start from the careful analysis and understanding of problems, and not leap straight away to what is in fact the second stage, the proposal of trial solutions.

In politics solution, real or attempted, are normally called policies. Every reputable political or social policy is a proposed solution to a problem; and we always need to be clear about the problem before we can propose the solution. We must always be able to ask of a policy: "to what problem is this the solution?". If there is no problem to which a given policy is a solution then the policy is superfluous, and therefore harmful, if only because it consumes resources to no purpose. Policies which are not solutions to any identifiable problem are part of the common currency of so-called practical affairs. Committees are especially good at producing them. I have stopped many a committee meeting dead in its tracks by asking the question: "To what problem is this the solution?". The whole notion that you can *start* with policies is deeply erroneous, and very damaging in practice. One of the forms it takes is starting from a blueprint, because of course a blueprint is a proposed solution; but it takes many other and more mundane forms. It is essential to start from *problems*, and to arrive at the formulation of each policy only as a solution to a problem.

According to Popper, in both politics and the growth of knowledge, criticism is the most effective agent of desirable change, and must therefore be not only free but welcomed, and acted upon. We can never be in a position to know that we have got things exactly right; our formulations and policies are always open to improvement; therefore any notions of certainty or unquestionable authority are not only out of place but damaging. The best we can *do*, like the best of our knowledge, is the best only for the time being, and in the prevailing circumstances. It is always, in principle, improvable, and therefore should always be subject to critical discussion.

In practice this attitude ought to breed a respect for political opponents, and a willingness to learn from them. In all the democracies I know, politicians lag behind the public on this matter. They would be more, not less, popular with their electorates if they were more willing than they are to admit error, and they would also be more, not less, popular if they were more willing than they are to admit that their opponents are quite often right.

The Popper approach constitutes a program for practical and rational improvement, and the usual word for that in politics is "reform": so it is a methodology of reform. But it leaves open the question of how quick or slow reform should be, the even more important question of how radical it should be, and the most important question of all, namely what it should consist of. This makes it an approach that can be adopted by anyone on the political spectrum between those who want no change at all and those who want revolution. What this means in practice is that it can be adopted by anyone committed to democratic politics: so it is also what you might call a methodology for democracy. It so happens that the youngish Karl Popper who wrote *The Open Society and Its Enemies* (1945) in the late 1930s and early 1940s had always been left of centre, and throughout the whole of his adult life up to that point a strongly, emotionally committed social democrat. But like so many people he moved to the right in middle age, and by the time of his death would have been accounted a conservative by most people – though to the end of his days he continued to regard himself as a liberal in the classic sense of the word, meaning someone who puts individual liberty first among the political values. My point is that his basic approach is one that can be adopted by anyone committed to democratic politics, from the extreme democratic left to the extreme democratic right, which indeed was the gamut that Popper himself passed through.

Having said that, though, the point has to be made that the Popperian approach sits most comfortably with a left-of-centre position, the sort of position Popper himself occupied when he produced it. This is because it gives rise naturally to a radical attitude towards institutions. It is not only policies that have to be seen as attempts to solve problems: institutions do too. A country's education system is its solution to the problem of how to educate its young; its armed forces are its solution to the problem of how to defend itself; its health services are the its solution to the problem of what public provision make for those of its citizens who need medical help; and so on and so forth. Just as in the case of policies, an institution that is not a solution to any problem is superfluous. Indeed, it is that condition that renders institutions obsolete. And because an institution is a practical solution to a problem, so long as it has a real function it is capable of being more effective or less, more satisfactory or less, more comprehensive or less, more expensive or less, more popular or less, and so on. The Popperian approach involves subjecting institutions to a permanently critical evaluation in order to monitor how well they are solving the problems they exist to solve – and involves moreover a permanent willingness to change them in the light of changing requirements. I have always taken the famous dictum of Jesus of Nazareth "the sabbath was made for man, and not man for the sabbath" to mean that we should bend institutions to fit human beings, not human beings to fit institutions; but this is at odds, I do believe, with some of the basic attitudes common to political conservatism, which include a reverence for institutions as such, a deep-seated unwillingness to change them, and a readiness rather to let their requirements override personal considerations. There is no logical incompatibility, but there is, I think, a certain psychological uncomfortableness in combining a Popperian approach to the requirements of institutional change with a typically conservative emotional attachment to existing institutions. The only kind of conservative with whom the two can sit comfortably together are those of the radical right, politicians like Margaret Thatcher, whose approach to traditional institutions was in fact highly disruptive.

The permanent monitoring of institutions to see if they are *not* performing as required, and the permanent monitoring of the implementation of policies to see if they are having undesirable consequences, are activities – and reflect a cast of mind – that come much more readily to radicals, of left and right, than they do to traditional conservatives. They also run counter to the way people working in institutions, especially those with authority, tend normally to behave. The normal tendency is to cover up organizational and administrative failures as much as possible, and to resist facing even to oneself the fact that one's activities are not having the desired effects. The Popperian approach, which requires one actively to seek out failures and shortcomings and do something about them, calls for a degree of intellectual honesty from politicians and administrators, as it does from scientists, that does not come to them at all easily, and constitutes a disconcerting personal challenge. What provides the incentive to meet this challenge is the higher success rate that results from doing so.

In fact a thoroughgoingly problem-solving approach has many practical advantages, perhaps even more in politics than in science. It is far easier to get agreement on problems than on solutions, and a government that starts from the problem – let us say, to take a small but emotive example, the problem of what to do about the number of homeless people sleeping rough on the streets of London – and then shows itself open to alternative possible solutions will probably have not only a higher degree of practical success than one that starts with the answer (in other words a policy), it will also enjoy more support and goodwill, even from those who disagree with what it eventually does. In a democracy a great deal of electoral advantage is to be had from a problem-solving approach, because people will feel that they have been brought in.

And of course, if I may be forgiven for stating the obvious, a problem-solving approach directs one's attention to problems, and makes doing something about them the first priority. It protects one from being seduced into trying to build utopia; and yet it does not easily allow one to relapse into complacency or inactivity. One's energies are channelled not into constructing ideal models but into removing avoidable evils. Popper encapsulates the first rule of thumb he recommends for public policy in the words "minimize avoidable suffering" (Popper 1945, vol. I, pp. 284-285). Psychologically it is a different approach from that of crusading for ideals, to which so many political activists are dedicated: it is more practical, and nearly always more fruitful. In any case the two are not necessarily incompatible. I am not opposed to idealists as such, but I do regard them with the gravest of suspicion. It is a fact that social evils have been perpetrated by idealists in our century on a simply stupendous scale that includes the deliberate murder of tens of millions of men and women and the herding of tens of millions more into forced labor camps (I am thinking not only of the Soviet Union but also of China, where the numbers involved may have been the greater). These things could not possibly have been done by people who had adopted "minimize avoidable suffering" as their guiding principle. But they were done by idealists, and condoned all over the world by other idealists, more often than not with a sense of moral self-righteousness accompanied by savage denunciations of anyone who criticized what they were defending.

A point Popper makes which I stress more than he does is the unavoidability of unintended consequences. I stress them because they often dominate practical politics – as they soon came to do in all communist societies, for example. An awareness of them also immunizes us against enthusiasm for any form of centralization, especially centralized planning. To anyone engaged in practical affairs, business as well as politics, they are of never-ceasing importance. Only on someone divorced from reality can they fail to impinge.

Political lessons to be learnt from Popper are not confined to the problem-solving approach and its method. He has certain large-scale perceptions about politics that seem to me right and important although unfashionable. For instance, he perceives clearly that the societies in which we, the West, are living in the 1990s are by all real (as against ideal) standards – that is to say by all the standards of past experience – exceptionally non-violent, as is the international scene as a whole. He also sees that for the great majority of men and women in the democratic West life is better now than it has ever been before, not only materially but in the most important non-material ways, for example health, education, and cultural opportunity. He therefore sees clearly that the cultural pessimism so fashionable today, when intellectuals and artists are saying on all sides that we live in a uniquely terrible and violent time, presents more or less the opposite of truth. I suspect that the illusion it represents has been brought about partially by the collapse of the historicist, progressivist illusions that were held earlier in the century by a great many of the same people, and to which Popper was equally opposed. On the face of it, it is peculiar that so many individuals who for decades believed with a kind of religious intensity that everything was getting better are now equally certain that everything is getting worse. But both attitudes are holistic and uncritical, and meet what seem to me primarily religious emotional needs. The fact is that the liberal democracies of the West are the only large societies in the whole of human history in which the great majority of the people have enjoyed not only material prosperity and literacy but also what have come to be known as fundamental human rights. This is a very recent historical phenomenon, and it is a wonderful thing. Even so, there is no contradiction at all between seeing this clearly for what it is and at the same time trying to improve these societies, and for that purpose adopting a radical and essentially critical stance in their political and social affairs. It happens to be the position I myself have always occupied, independently of Popper, and it is what first drew me to his work, before I knew anything about his epistemology or his philosophy of science.

Another overall perception of Popper's which I share is that equality of outcomes is not a desirable social goal. It took me a long time to learn this lesson, and when I did it was not from Popper but from my poor constituents in east London. They were almost entirely without social envy, which I came through them to realize is a largely middle class phenomenon anyway. They wanted a better deal for themselves – better wages, better houses, better schools for their children, and so on – but had no desire to pull down anyone who was better off. On the contrary, they actively rejected any such attitude; it ran counter to some of their most basic aspirations, more often for their children than for themselves. And they saw it as incompatible with elementary personal freedom. They were right in this. And it was also Popper's view. He once said that if a form of socialism could have been discovered which was compatible with personal freedom he would still be a socialist.

Another general attitude of Popper's that I loudly applaud is his hostility to the tyranny of fashion in all its forms – the idea that we have to do certain things, or do things in certain ways, because these are the 1990s, and that we really have no choice, in that anything else is contrary to the spirit of the times, and therefore inappropriate, perhaps even inauthentic. This error is at its most predominant and destructive in the world of the arts, but it operates in politics too. In Britain after World War Two we had years of uncritical commitment to Keynesian economic management followed by uncritical commitment to monetarism; we had an uncritical belief in nationalization followed by an uncritical belief in privatization. Town planners guided by what they took to be the spirit of the times devastated the centers of many of Britain's most beautiful towns during the 1960s and 1970s, and corralled the poor of the inner cities into tower blocks. Anyone who opposed these developments at the time was denounced as conservative or reactionary, fuddy-duddy, out of date. Popper has always believed in either fighting or ignoring such tides of opinion. He sees them as forms of what another kind of philosopher would call "false consciousness," and as ways of evading responsibility for our own decisions and our own actions. Insofar as we go along with them we are enemies of our own freedom. We can do *whatever* we can do, and it is up to us to do the best we can.

One of Popper's specific proposals that I think has great merit is that it should be accepted internationally as a fundamental principle that no existing frontier is to be changed except by peaceful negotiation. The point here is that nearly all the national frontiers in the world were established by force, usually either imposed on the vanquished by the victors in war or imposed on colonized peoples by imperialist powers; therefore if the fact that a frontier has been imposed without the consent of one of the parties is to be accepted as an excuse for that party to use violence to get it changed, there would be justified wars breaking out all over the world all the time, several on each continent. This cannot be acceptable to the international community now. Existing frontiers, constituting as they do actually existing political reality, must be regarded by the United Nations as operative no matter how they were arrived at, and must be guaranteed by whatever international peace keeping forces there are, unless a majority of those whose frontiers they are wish to change them by peaceful means.

Up to this point I have been endorsing Popper's approach and commending it to you. And the truth is I do believe it provides working politicians with rules of thumb of the utmost usefulness. But it does have, inevitably, limitations and shortcomings. The chief limitation is that, being a methodology, it is almost entirely about method and not about content. But the most pressing question facing the individuals who have to take important decisions is nearly always "What should we do now?". Everyone else can stand back from that question and then criticize the way things are done, but the decision-makers themselves cannot. Only rarely does the Popperian approach help them towards an answer. This fact has recently come to the fore in the former communist countries of Eastern Europe and the Soviet Union. To an extent rare in history they have found themselves with opportunities to build a new society that is radically different from the one they had before. Popper's philosophy offers them first-rate guidance about how to do things, but very little about what to do. What kind of local government, if any, do they want: at what level, how constituted, and with what powers? What kind of education system do they want, what sort of schools, how organized, by whom, teaching what? How much welfare state do they want, and in what areas – and how much can they actually afford: how is it to be administered, how funded? It is questions like these that constitute most of the content of large-scale practical politics.

And in any case, most politics is not large-scale. When I became a Member of Parliament and began spending my days in the House of Commons among hundreds of other MPs, I was much struck by the fact that, among themselves, they scarcely ever discussed the sort of political or social questions thrashed out in pubs and debating societies, like are we in favor of the return of the death penalty, or censorship, or nationalization. The questions that held them in thrall were much more like: "If we raise the widow's pension by half a percentage point, where are we going to find those extra millions of pounds?". They would have differing views about such questions, and would argue heatedly, but these mostly were the sorts of questions they would be arguing about. And it is inevitable that these are the sorts of questions that day-to-day government has to concern itself with. It is seldom that Popper's work offers much guidance with them.

This in itself is not a criticism of Popper, because he is not talking to us on that level. From a philosopher a politician must expect strategic, not pragmatic, guidance. What I am drawing attention to is not a shortcoming but a limitation. It is, however, one that practical politicians are likely to be a lot more conscious of than other people.

Practical politicians are only for a very small part of the time concerned with putting principles into practice. Most of the time they are struggling to make the best they can of difficult, messy and uncontrollable situations. I will give you an example of this that involves a conflict between me and Popper personally. I have already mentioned his conviction that the international community should impose an iron refusal to allow existing frontiers to be changed by force, and have given his reasons for it. Well, when the Military Junta then governing Argentina invaded the Falkland Islands, for which Britain was responsible in international law, and *de facto* war began, he telephoned me at the House of Commons in great passion, wanting me to urge the British government to declare war formally on Argentina. I refused. What

I said to him went roughly as follows. "I agree that the Argentineans absolutely must be made to leave, by negotiation if possible but by force if necessary. And I will vote for the use of force if there is no other way. But I want to get them out with the minimum possible harm to everyone concerned, and I see this as a damage-limitation exercise. There happens to be a sizeable British community living permanently in Argentina that consists of tens of thousands of families, many of whom have been there since the 19th century. They have their own schools and other institutions, as well as their own homes, businesses and professional practices. If we declare war on Argentina, the Argentinean government may well intern them and confiscate their assets. Their whole world will be destroyed, and in many cases their individual lives will be ruined. I believe we can get the Argentineans out of the Falkland Islands without that happening – though only if we don't declare war."

Popper, always willing to sacrifice himself to a principle, was willing to sacrifice others too, and would not agree with me. Not only did he continue to telephone me angrily throughout the Falklands war, always urging the same course of action on me; he continued to bring the subject up with me for the rest of his life, always maintaining that he had been right. I am convinced to this day that he was wrong – and not only because what I wanted to happen did in fact happen. I fully acknowledge that it might not have done. But I am convinced that we were right to try. I re-emphasize that I was always completely in agreement with Popper that in no circumstances should Argentina be allowed to get away with the forcible annexation of the Falkland Islands. He and I differed only about how they were to be made to leave. But on this we differed profoundly. It was not the principle that was in dispute but the way it should be put into practice. Popper wanted commitment to the principle to be publicly proclaimed in a formal act: I saw this as unnecessary to the actual implementation of the principle and almost bound to be seriously damaging. So I saw my own approach as essentially practical and his as essentially theoretical – but far too theoretical, culpably so, too little concerned with the actual lives of individual men, women and children. And I have to say, as an intellectual and academic myself, that I see this fault as all-pervading in the attitudes of intellectuals and academics to political and social matters, and as being an extremely serious, often debilitating fault. Also, having been a professional politician as well, I find the sense of personal superiority to politicians so commonly expressed by intellectuals and academics unfounded and misplaced, self-deluding.

This story of a clash between a political philosopher and a professional politician illustrates a point of profoundest importance. I do not believe that there are many people who hold Popper and his work in higher regard than I do; and I knew him well personally. As a professional politician I made conscious use of his methodology, and found it of extraordinary practical usefulness and fruitfulness. Yet any individual who, if only by his vote in an assembly, has to take responsibility for executive political decisions, is likely to find himself unable to put Popper's principles – or anybody else's principles, for that matter – into practice in a way that the originator of the principles would wholly approve of. This is because practice has unavoidable and compelling exigencies which theory can never encompass, and which those who are solely theoreticians seem only rarely to appreciate – and never fully to understand. But that would be a subject for a different lecture.

University of Oxford
Wolfson College
Linton Road Oxford OX2 6UD
United Kingdom

REFERENCES

Popper, K.R. (1945). *The Open Society and Its Enemies*. London: Routledge.

Enrique Suárez-Iñiguez

KARL POPPER'S EDUCATIONAL CONCEPTION

Karl Popper is one of the most famous philosophers of the 20th century, nevertheless, as he himself has pointed out on several occasions and as has been noted by his main followers (see Popper 1976; Magee 1985; Bartley III 1982), Popper has not been well understood. In reality, he has not been well nor sufficiently read. The Popperian myth, as Popper himself has baptized it, has been attributed more importance than his true ideas. In part this is due to what famous people have said about him and that has often been repeated, even when not true, and in part to the tedious and repetitive nature of reading his works in complete form, which dissuades many from undertaking this arduous task. However, and in spite of this, Popper's works are full of thought-provoking ideas; not only the central ideas he is famous for – open society; the method of piecemeal engineering; his criticism of historicism in general, and of Plato, Marx and Hegel in particular; his proposal for using the deductive methods of testing and refutation as solutions to Hume's and demarcation problems, respectively; his theory of three worlds; Popper's Darwinism, etc. – but also from ideas sketched out here and there but not fully developed. One of these ideas is his educational conception.

Ever since his youth, Popper had been interested in educational matters. In 1925, as a student, he worked in the recently founded Pedagogical Institute of Vienna. At that time he wrote articles in educational publications but then seemed to forget about the topic. However, I suggest that behind many of his philosophical ideas there is a conception of how we should educate youth so they can have a place and participation in the open society he defends. Not just that but, in a certain way, his ideas on education are the *crowning* touch of his political and social philosophy. It is no coincidence that they are right at the end of *The Open Society and Its Enemies.* I know of no other author who has

In: E. Suárez-Iñiguez (ed.), *The Power of Argumentation (Poznań Studies in the Philosophy of the Sciences and the Humanities,* vol. 93), pp. 71-78. Amsterdam/New York, NY: Rodopi, 2007.

reflected on this point. Popper's philosophy of science also implies important educational notions.

The first thing to remember is his concept of an open society vis-à-vis a closed one. The latter is a tribal, magical society, in which there is no distinction between the duality of facts and standards, in which there is no freedom or democracy. Closed societies are ritual ones, as Jacob Bronowski understood so well. Easter Island, with its colossal statues, magnificent but similar and repetitive, with their empty eye sockets, are the expression of a closed society (see Bronowski 1973). But closed societies may also be modern ones. Totalitarian regimes are a clear example of such: there are no political liberties nor do social institutions have the needed weight. An open society, in contrast, is based on the dualism of standards and facts[1] and on the values of freedom, equality, humanity and reasonableness. And these are the first elements to be sought by any education whatsoever in a modern open society: encouraging respect for the values liberalism and democracy bring with them, freedom and equality. And as Tocqueville so perfectly understood, these two terms should go hand in hand. The best guarantee against the tyranny of the majority will always be freedom. Teaching children to love freedom as a good without equal and encouraging respect for certain egalitarian values, without falling into excesses, are central educational functions in a modern democratic society. Civic values are rooted in them. The two other values I mentioned before, humanity and reasonableness, likewise go hand in hand. Believing in man, with his dreams, goals, expectations, achievements and failures, knowing we can learn from our mistakes, that it is in our hands to progress and to use rational criticism as a means to correct our errors are important aspects of our learning. Man, a feasible being, is, however, the center of the world among other things, because of his capacity to utilize reason. And as in

---

[1] Dualism of facts and standards is one of the bases of the liberal tradition. According to Tarski, a statement is true if it corresponds to the facts. Facts and standards are different things. But truth, corresponding to the facts, can, in turn, be a moral standard and be valid within the realm of standards. In both realms we can learn from our mistakes and make our knowledge grow, but we should distinguish one from the other. Liberalism places emphasis on the improvement of our standards, specially in the area of politics and legislation. See "Facts, Standards, and Truth: A Further Criticism of Relativism," *addendum* to the 4th English edition of vol. II of *Open Society*. We cannot, as Lord Boyle said paraphrasing Popper, derive our standards and moral decisions from facts: a moral decision is taken. Of course, it can be based on a fact (all men are not equal) and, based on that fact, set the standard. As Popper says: we standardize in order to have equal rights. See Boyle (1974, pp. 843-858) and Popper's reply (1974, pp. 1153-1158).

the classical thought, the control of ourselves means the command of reason upon emotions.[2]

To Popper politics has a clear ethical dimension. If for ancient philosophers the objective of politics was to make citizens better and more virtuous, and happier as a result, for Popper the role of politics is to solve serious social problems. He does not believe that politics can make individuals happy, since he considers this a personal task,[3] but instead that politics should avoid or alleviate human suffering: "dreams of beauty have to submit to the necessity of helping men in distress, and men who suffer injustice, and to the necessity of constructing institutions to serve such purposes" (Popper 1945, vol. II, p. 165). Therefore, politicians should stop proposing utopic goals and devote their efforts to detecting and solving problems. Their role is to accept errors, not to cover them up.[4] We know that in politics this means the formulation, application and follow-up (with corrections) of policies; it means the creation and perfection of solid and fair social institutions. This is why the education required by an open society should necessarily lead to respect institutions and to the fostering of social and political participation in them as essential conditions for an ordered and democratic life. Without institutional life there is no open society. Without the existence of institutions, the application of law is impossible. Now then, modern open society is a democracy and "only democracy provides an institutional framework that permits reform without violence, and so the use of reason in political matters" (Popper 1945, vol. I, p. 4).

Popper's criticism of historicism, which encompasses holism and essentialism, is internationally famous. He has clearly shown that there are no historical-social laws that predict the future, that social sciences cannot formulate long term historical prophecies, and that the future depends on us and not on any historical necessity. As a defender of methodological individualism, Popper sustained that all social phenomena, especially institutional functioning, should be considered the result of decisions, interests and actions of individuals, and not the result

---

[2] We have to based on clear thought and experience, rather than emotions and passions. We should try to make reason "play as large a part as it possibly can . . . " See *Open Society and Its Enemies* (Popper 1945, vol. II, p. 224).

[3] Lord Boyle was right on target when he criticized Popper for this standpoint. According to Boyle the possibilities of being happy can also be increased. Boyle (1974, p. 855).

[4] "To look out for these mistakes, to find them, to bring them into the open, to analyze them, and to learn from them, this is what a scientific politician as well as a political scientist must do" (Popper 1961, p. 88).

of collectives ones. He was a lucid opponent of essentialist doctrines. All this implies an underlying educational notion.

Popper has understood, as no one else, the damage caused by these doctrines and has alerted us to them. He knew that if philosophy had the duty of unmasking false doctrines, education has to fight against them on a daily basis. Popper's struggle against historicism (and against holism and essentialism) is not only in the terrains of philosophy and politics, but in ethics as well. It is necessary not only to understand the reasons why these doctrines are harmful and mistaken, but to uproot them from the human being and to provide better and more useful theories in exchange. If we want to live in an open society, then it is necessary to educate the new generations with respect to better attitudes and ways of thinking. It is not possible to hope for a better world based on ankylosed forms.

The relationship between politics, ethics, and education comes down to us from the Greeks. Although Aristotle knew that ethics and politics are practical, i.e., one learns by doing, he stated that the practice follows after having learned the fundamentals. He thought that the role of political science was to teach citizens to be virtuous, just as the role of ethics was applied to personal lives. Together, ethics and politics formed "the philosophy of human things." That is the reason why his *Politika* begins where his *Nichomachean Ethics* ends. To Plato educating is forming virtue. For Greeks education is the link that joins ethics and politics (Suárez-Iñiguez 1993; 1996). In this sense Popper's philosophy recovers the Greek postulates and is an attempt to place again a dyke – the dyke of ethics – to politics, without which there would only be power in the hands of a few to the detriment of the rest. But, at the same time, it places modern liberties above ancient ones, as Rawls (1982) would say, i.e., it places the individual above the collectivity and in this sense it is not Greek at all. Popper is a liberal because he recognizes the value of individual liberty and is sensitive to the dangers inherent in all forms of power and authority (Popper 1963). In synthesis, the creation and a good working order in an open society is not only a political concern or, better stated, it is such if we conceive politics as the other part of the philosophy of human things.

Now, in all ethical and political philosophy, how to attain what one wants to achieve is fundamental. Popper made a devastating criticism of the method of "wipe the slate clean," which unfortunately is deeply rooted and that, according to him, can be found in Plato's *Republic.* This method is dangerous for it requires attempts that will prove to be unfruitful since never, or at least almost never, do things turn out well after the first attempt. In exchange Popper proposed what he called the method of piecemeal engineering, which implies learning from experience, by trial and error and through small gradual adjustments. The difference between the piecemeal method and the essentialist one is huge:

> It is the difference between a reasonable method of improving the lot of man, and a method which, if really tried, may easily lead to an intolerable increase in human suffering. It is the difference between a method which can be applied at any moment, and a method whose advocacy may easily become a means of continually postponing action until a later date, when conditions are more favourable. And it is also the difference between the only method of improving matters which has so fare been really successful, at any time, and in any place . . . and a method which, wherever it has been tried, has led only to the use of violence in place of reason, and if not to its own abandonment, at any rate, to that of its original blueprint. (Popper 1945, vol. I, p. 158)

I do not have to say that these ideas are valid both for individual life as well as for social life. They are two radically different ways of conceiving and living life: one, useful, susceptible to daily improvement; the other, sterile, sacrificing the best of life, the present, for the benefit of utopian goals; one, seeking adjustments; the other, inquiring into essences; one, that is present-day; the other, worried about origins. Citizens of an open society should be educated in the piecemeal method based on reason and whose values are humanism, individualism, freedom and equality: on respect for and recognition of the importance of institutions and divested of essentialist, holistic and historicist notions and on the concept of "wipe the slate clean."

As I stated at the beginning, it is not only Popper's political philosophy that rests on an educational conception. His philosophy of science includes it too. According to this, Popper sustains that the advancement of knowledge in general and of scientific knowledge in particular, is achieved through conjectures and refutations, and that refutations should be sought by means of testing our theories, which in turn can be achieved through observations, experiments and rational criticism, among other things. If the tests reject the theory, this simply means that the theory was false. With his now famous example of the white swans, Popper showed that science proceeds deductively and that it is impossible to prove; one can only refute universal theories.

From here stem fundamental educational notions. The most obvious one is that we should not wed ourselves to our own ideas, but instead should be open to rational criticism. Besides, since we cannot prove any hypothesis but only corroborate it through significant tests and only for the time being, we should not be vain nor try to know more than everybody else. According to Sir Karl, ideas are conjectures and the important thing is not to prove a particular conjecture but to advance in our knowledge. Refutations precisely allow us to discover errors and thus to know more, so they should be welcome. In other words, Popper taught us that our knowledge is very limited and that we should be humble in our honest search for the truth. He provides us with guidelines to discern whether or not one theory is better than another, and if we are closer to or farther from the truth. All this implies a radically different approach to scientific research and to life itself.

Now, if fighting mistaken and harmful ways of thinking and proposing others which are better is not an educational task, what could we understand an educational task to be? Combating essentialism and its questions of the type "what is it?" and "when did it happen?"; questioning the holism and emphasizing the importance of the individual; criticizing behaviorism, among other reasons, for its simplistic viewpoint, and psychoanalysis, for non-scientific; denying the existence of unavoidable historical laws and the validity of prophecies in the social sciences (historicism); arguing that we should not sacrifice the present in the name of an ideal future and that we all have the right to live and to seek weel-being; showing that the theory of the "wiping the slate clean" is dangerous and, in particular, harmful because useless; teaching that there is no sense in collecting data if we have no prior theory; sustaining that we can never demonstrate a theory conclusively but only corroborate it; insisting that we learn by trial and error and by piecemeal adjustments; maintaining that we should be responsible in our decision-making and foresee its consequences; denying the existence of "keys" to understand history and asserting that it is we who make history, all this forms part of an educational notion that plays a central role in Popper's philosophy; moreover, Popper believes that these theories can be taught.

We try to learn, Sir Karl has said, that what judges us is our conscience and not our success in life and that it is possible even to contempt power, glory and wealth so as to act according to our conscience, fulfilling our duty the best we can. The problem with education today is that "we are educated to act with and eye to the gallery." We live for others and for false social values. It is an ethics of fame and fate. If, instead of that, we opt for acquiring a relative appreciation of our own importance vis-à-vis others and for fulfilling our duty by finding satisfaction in doing our work without expecting praise or absence of blame, then, and only then, he states, can we be the responsible and well-educated individuals we ought to be. We need an ethic that disdains success and reward. We need dare to be free. Popper believes deeply in the value of the individual and his transforming capacity and thereby adopts the Kantian conception of the autonomy of will as the supreme principle of morality.[5]

As Cassirer said, paraphrasing Kant, we should think, judge and decide for ourselves or, as Thomas Alva Edison once said "anything that *man's* mind can invent, man's character can control." Educating, for Popper, is teaching the use of one's own reason so as to know what is wrong and mistaken, correct the way and elect the best one, intellectually as well as morally and socially.

The world of today seems not to favor these educational concepts. It impregnated, as it is, with the ethic of fame and of fate, in favour of superfluous values and doubtful ends. And notwithstanding, it is the only way of saving ourselves. Virtue, Thomas More said, is the life oriented according to nature and follows the course of nature those who are governed by reason. Socrates, Plato, Aristotle, Cicero, and Seneca said something similar. Indeed, education should be based on the ancient principle of doing no harm and doing all the good possible. Just like Saint Augustine and Saint Thomas, Popper believes this is fundamental; the need to apply the Golden Rule: "do onto others as you would have them do unto you." Man has to believe in something more than just in himself: something to fight for and to sacrifice for, though it is certainly essential first to believe in himself.

Popper believes we should give youth what it most urgently needs to become independent of us and be able to choose for itself. We have to value present life and learn to make adjustments that correct errors and to solve concrete problems. Once again, we should disdain eloquent sounding goals, and fight to suppress avoidable pain and to alleviate the unavoidable insofar as we can. This is the educational legacy that Popper left us.

Universidad Nacional Autónoma de México
Facultad de Ciencias Políticas y Sociales
04510 México
Distrito Federal
México
e-mail: esiiguez2002@yahoo.com.mx

---

[5] On the Kantian roots of Popper's political thought, see Ryan (1985).

REFERENCES

Aristotle (1983). *Nichomachean Ethics*. Translation by A.G. Robledo. Mexico: (Bibliotheca Scriptorum Graecorum et Romanorum Mexicana) Universidad Nacional Autónoma de México.

Bartley, W.W. III (1982). A Popperian Harvest. In: Paul Levinson (ed.), *In Pursuit of Truth. Essays on the Philosophy of Karl Popper on the Occasion of his 80th Birthday*. pp. 249-283. Atlantic Highlands, NJ: Humanities Press, Inc.

Boyle, E. (1974). Karl Popper's *Open Society*: A Personal Appreciation. In: P.A Schilpp (ed.) *The Philosophy of Karl Popper*. La Salle, IL: Open Court Publishers.

Bronowski, J. (1973). *The Ascent of Man*. London, British Broadcasting Corporation.

Magee, B. (1985). *Philosophy and the Real World. An Introduction to Karl Popper*. La Salle, IL: Open Court.

Popper, K.R. (1945). *The Open Society and Its Enemies*, vol. I. London: George Routledge and Sons.

Popper, K.R. (1945). *The Open Society and Its Enemies*, vol. II. London: George Routledge and Sons.

Popper, K.R. (1961). *The Poverty of Historicism*. New York, London: Harper Torchbooks, Harper and Row Publishers.

Popper, K.R. (1963). *Conjectures and Refutations*. London: Routledge and Kegan Paul.

Popper, K.R. (1976). *Unended Quest: An Intellectual Autobiography*. La Salle, IL: Open Court.

Rawls, J. (1982). *The Basic Liberties and Their Priority. Tanner Lectures on Human Values, Michigan University*. Salt Lake City, UT: University of Utah Press and Cambridge University Press.

Ryan, A. (1985). Popper and Liberalism. In: G. Currie and A. Musgrave (eds.), *Popper and the Human Sciences*, pp. 89-103. Boston, Lancaster: Martinus Nijhoff Publishers.

Suárez-Iñiguez, E. (1993). *De los clásicos políticos*. Mexico: Miguel Ángel Porrúa-FCPyS Universidad Nacional Autónoma de México.

Suárez-Iñiguez, E. (1996). Las ideas políticas de Platón. *Estudios Políticos* 12, 89-113.

PART II

PHILOSOPHY OF SCIENCE

Joseph Agassi

CORROBORATION SPURIOUS AND GENUINE

1. The Problem

To put Karl Popper's theory of scientific research the way Professor
Jarvie did earlier in his chapter, science is a Socratic dialogue. In it
ammunitions are the *general* facts that Michael Faraday and Claude
Bernard called experimental arguments.[1]

Progress in science is the development of new ideas and the finding of
new *general* facts. New ideas are imagined; they are conjectures. How
are new *general* facts found? By attempts at criticism, says Popper, by
attempts to refute our conjectures.[2] This is an intellectual venture. The

---

[1] Faraday (1939, paragraphs 1799 and 2010). See also my *Faraday as a Natural
Philosopher* (1971, pp. 64, 132, 137, 176, 295), Bernard (1940). On page 28, Bernard
begins refering to "a priori ideas" as "experimental ideas." His main term becomes
'experimental ideas'. He says:

> The metaphysician, the scholastic, and the experimenter all work with an a priori idea.
> The difference is that the scholastic imposes his ideas as an absolute truth which he
> has found and from which he deduces consequences by logic alone. The more modest
> experimenter, on the other hand, states an idea as a question, as an interpretative,
> more or less probable anticipation of nature, from which he logically deduces
> consequences which, moment by moment, he confronts with reality by means of
> experiment. He advances, thus, from partial to more general truths, but without ever
> daring to assert that he has grasped the absolute truth. (Bernard 1940, p 27)

The term 'anticipations of nature' is Bacon's. The term 'Interpretative' alludes to William
Whewell. The term 'the absolute truth' should read, 'the whole truth', of course, a point
stressed by his most celebrated follower, Pierre Duhem.

[2] The *locus classicus* for Popper's view that proper corroboration is attempted refutation is
his *The Logic of Scientific Discovery*, Chapter 10 and Appendix * IX. For further detail,
see there, Index, Art. *Corroboration*. For my own view of the matter, which differs from
Popper's on points not relevant to this essay, see his *Conjectures and Refutations*. See
also my *Science in Flux*, Boston Studies, 1975, Appendix to Chapter 1, "The Role of

In: E. Suárez-Iñiguez (ed.), *The Power of Argumentation* (*Poznań Studies in the Philosophy
of the Sciences and the Humanities*, vol. 93), pp. 81-100. Amsterdam/New York, NY:
Rodopi, 2007.

prize for success in this venture is learning from experience; corroboration is a consolation prize for its failure: if a hypothesis is false, then we learn more from experience by refuting it than by corroborating it, but this is not up to us: the proper experimental setting is the way to make nature decide the outcome, since, if the outcome of an experiment is known in advance, then there is no need to perform it.

An empirical claim that conflicts with this theory is that scientific evidence is found in the search for corroborations. Does this empirical claim refute Popper's theory empirically? The answer received among philosophers of science is in the affirmative. And so, Adolf Grünbaum had the insolence to declare Popper a myth-maker. For my part, I take the empirical claim in question seriously because of my profound reluctance to accommodate criticism *ad hoc*. Am I defensive about Popper's theory unawares? My refusal to admit the empirical claim in question as a refutations of Popper's theory, should it compel me to devise an *ad hoc* escape measure? Heaven forbid. I consider it both irrelevant to Popper's theory and false. It is irrelevant to Popper's theory since that theory pertains only to the search for knowledge without *ad hoc* measures. If anyone will contest my presentation of Popper's view, then I will claim it as mine; perhaps it is Popper *à la* Agassi; surely it is at the very least a significant variant of Popper's theory (see note 2). On this theory, is the search for corroboration a search for knowledge? If and to the extent that corroboration advances science, then, yes; otherwise, not. What, then, is the capacity of corroboration to advance science? Since the claim in question pertains to research activities, we should preface this question with the following: how do private and public aims interact within science? How do private and public aims interact at all?

2. Ends Private and Public

The aims of science, if these exist, must differ from the aims of individual researchers, since individuals have diverse private aim. Moreover, the aims of science are good, yet aims of individual researchers are occasionally bad. The recognition of the difference between the private and the public leads to the question, what abstract entity is science that it should possess aims? More generally, public,

Corroboration in Popper's Philosophy," and Chapter 13 and 14 there on positive evidence. See also there, Chapter 12, "The Confusion Between Science and Technology in Standard Philosophies of Science."

abstract entities, institutions or other, how do they possess aims in the first place? Obviously, the public ends are *unlike* individual ends; individuals recognize them as public or institutional.[3] Thus, we find implementations of the aims of science almost everywhere – and of those of communism almost nowhere. This point was elaborated by Fred Eidlin in his chapter. Rational processes, then, depend not only on institutional ends, but also on ends of individuals acting on its behalf and on the consistency or otherwise of institutional and individual aims. Consider an individual whose aim is utterly personal, and suppose that this personal aim can be achieved only by furthering the aims of science. Then, says Popper, the personal aim drops out of the picture. Instead of the question of individual aims, the question can be raised how come the institutional end is what it is, how come sufficiently many individuals endorse it, how is it guarded. How come most researchers are able and/or willing to try to enrich human knowledge rather than their own purses?

The aims of a researcher may be the discovery of a truth, the corroboration of a conjecture, the completion of some technological task, and the desire to win immortality, an academic appointment or a prize. Discussing the may in which these aims interact is enlightening, as they can support each other, or be in conflict. (Compare two individuals who wish to reap the reward which the advancement of science may earn them, one doing so by honest toil, the other by theft.) Similarly, at times the desire to have one's conjecture corroborated is in accord or in conflict with the wish to discover the truth, as the corroborated conjecture may be true or at least on the right track, but it may also be spurious and misleading.[4]

3. Success Spiritual and Worldly

Success is the achievement of aims, individual or institutional. In addition, there is the institutional recognition of individual or of institutional success. One major institutional recognition of individual success is that of individual worldly success. The view of individual

---

[3] On institutions and their aims see my "Methodological and Institutional Individualism," reissued in Agassi and Jarvie (1987).

[4] For an interesting example of the inability to cope with a misleading corroboration see Mackie (1936-1937, pp. 55-67), and my comments in my *Towards an Historiography of Science, History and Theory* ([1963] 1967, p. 44 and Note 122). See also José Romo and Manuel G. Doncel (1994) and the Manuscript of Series 1 of his *Researches, Arch. for Hist. of Exact Sciences* **47** (1994), 291-386.

success as worldly is almost always correct, since most people desire worldly success. Nevertheless, no matter how few are the individuals whose desire are uncommon, this forces us to judge individual success personally: one succeeds if and only if one achieves one's own aims, whatever these are. We may describe the success of an action, then, once as worldly and once as relative to personal desires. This, incidentally, holds equally for good aims as for bad ones: good or evil people succeed only when they achieve their own respective good or evil aims. This leaves open the question, is the aim of having one's theories corroborated good and if so why? Or is it bad, and if so why?

Consider the aims of researchers. Let us divide them to the worldly and the spiritual – and let us view intellectual pleasure as more spiritual than worldly. In societies that offer worldly rewards for spiritual success this pleasure is in the overlap between them. Here our concern is with the advancement of science, not of profit, with scientific success as spiritual, with action for the greater glory of God, as the expression goes. Can we argue that success here require corroboration? Yes, Popper says, reports of *general* facts must be corroborated. This is an important, and it is often ignored in the presentation of the discussion, only to be dragged in as an argument, seemingly against Popper's thought. Significantly, in generalizing science (by definition) only general facts are considered, never the singular facts that may enter the historical sciences (political history, the history of science, geology, etc.).[5] This is the only rigid rule that was never controverted in the generalizing sciences: singular observations are not to be regarded except if they are generalized. It was made by Galileo and by Descartes; Robert Boyle instituted it in the Royal Society of London in the 1660s.[6] Whereas most of the philosophers of science still discuss the way singular observations support their generalizations, Popper is the only one who discusses the situation in details that agree with the scientific tradition: only testable statements are

---

[5] Hegel and his followers, including Marx, declared that the human sciences are historical, but not the natural sciences. The refutation of this, rather naive view was stated clearly by Wilhelm Rickert, who observed that there are historical studies of nature and general laws of human matters. The latest significant discussion of it is in Karl Popper's *The Poverty of Historicism* (1957) and in I.C. Jarvie's comments on it (Jarvie 1982).

[6] The view of the matter was presented in the Proëmial Essay to Robert Boyle's *Certain Physiological Essays* of 1661. The rules it prescribes were endorsed by the Society soon afterwards, in a vote on a motion made by its president and seconded by Boyle. The rules have repeatedly received significance in recent decades, with the reports on single experiment that take a relatively long time – weeks or months – to repeat during which the need to suspend judgment on it is often expressed.

scientific, and so the scientific statements of observed fact are thus corroborated generalizations. They are accepted not as foundations but as challenges, as items to be explained. Since we are concerned here only with theories, not with factual reports, let us not consider this a point in favor of Popper's theory and recognize, as a generally admitted fact, that both general facts and their corollaries are in need of corroboration. Incidentally, an important category of consequences from generalizations that thus require corroboration is that of purely existential statements: electrons exist; quasars exist; democratic abdications of democracy exist. These seem to be theoretical statements, as they employ theoretical terms, so called. The discussion of theoretical terms, however, is illusory, since, anyway, *all* observation reports have to be theory-laden, as they *all* employ theoretical terms, since, as Popper has observed, *all* nouns are dispositional.[7]

So let us return to the dispute at hand, the dispute that concerns theories, on the supposition that they are conjectures. When is a conjecture a success? Can we argue that the success of a theory require corroboration? If yes, then, are there instances for this? If yes, then there is a refutation of the view, Popper's or mine, that research is conjectures and refutations *sans ad hoc* measures.

There are repeated observations on record of the joy of scientific discovery, and these are made at times in terms more worldly than spiritual, and at times the other way around. Even researchers who enjoy their finds both materially and spiritually, may differ in the way they enjoy them – when they view their contributions as primarily personal achievements or as primarily contributions to the advancement of science. The difference expressed in many ways, at times significant.

To take one specific significant example, a few researchers have refused to patent their findings, including prosperous Robert Boyle and struggling Michael Faraday. Their refusal to patent was a matter not of principle but of a social gesture: they intended to accentuate publicly that the benefit of scientific research is primarily spiritual. The examples to the contrary are honorable too: researchers who enjoy their own finds more materially than spiritually are still contributors to human understanding and/or welfare. Regrettably, though, their enjoyment is not as full as possible, so that on this they are not role models. I met some of them in person and found their sense of science very different from the

---

[7] See Popper (*The Logic of Scientific Discovery* 1959, end of Section 25). For further detail, see there, Index, Art. *Disposition*. The tremendously large and futile literature on dispositional terms, that began with Carnap's *Testability and Meaning* of 1936, stems from this discussion of Popper.

Joseph Agassi

one described by Popper. So, understandably, I was delighted to read the following in David Budworth's (1981) autobiography, *Public Science Private Views*: Popper gives him the sense of regret that he has deserted activity in scientific research for the sake of administrating it, and Thomas Kuhn effects him the opposite way.[8]

4. The Aim of Scientific Research

The wish to enrich the fund of human experience regularly involves the search for means useful for this end, and if the only means for that is refutations, then those who wish to learn from experience do seek refutations however reluctantly: desired ends justify distasteful means (unless these are forbidden). If Popper's theory is true, then researchers who report that they seek corroborations are mistaken – either about their aims or about the means for reaching them. Attempting to test this corollary empirically, one may request researchers to report their preference between corroboration and progress – on the supposition that they cannot have both. This test will fail, as they might refuse this very supposition. To have the test succeed, then, one has to devise a non-suggestive questionnaire, which is hard to do, especially when the group to be studied is of sophisticated researchers.

For my part, I prefer to avoid discussing private aims; my personal preference is to begin this discussion with the question, when and how does corroboration advance our knowledge of the world? For, corroboration may have different roles in different situations, especially in situation governed by different aims, especially in technology, where individual aims differ, so that criteria for success differ too. This, however, is outside the problem-situation as presented here, since we are here discussing scientific research. Nevertheless, it may be observed, perhaps, Popper's view of corroboration fits the facts about corroboration in technology better than any other theory: in technology attempts at refutations are more glaring than in science, since in modern societies they are prescribed this way by the law of the land.[9] Hence, though in technology the aim of those who pay for a test may indeed be to corroborate their wishes, nevertheless, even there tests are performed as attempts to refute, despite the hope to corroborate. This, of course,

---

[8] "In passing I cannot resist remarking that, while reading Popper makes me wish I were still an active scientist, reading Kuhn makes me glad that I am not" (Budworth 1981, p. 177).

[9] See note 2, above. See also my *Technology: Philosophical and Social Aspects* (1985).

invites cheating, and so technological tests are publicly controlled. But even otherwise, test with the wish to refute do take place, as these are performed by opponents or competitors: the implementation of any innovation may be desired by some and disliked by others. As Professor Magee has pointed out in his chapter, this is particularly true for politics, and it renders public criticism there essential.[10] Query: can that be the case in science? Are there parties within science? Yes, indeed. Can one scientific party wish to have a hypothesis refuted and another corroborated? Not if they share aims: in that case they all hope that the joint aim be furthered.

What then is the case "in real life"? Even though all researchers share the aims of science, we are informed, competitors wish their own views to be corroborated. Query: does that wish interfere with the result? We think not: we observe that this translates into increased rigor, in increase of the severity of the tests. Faraday, and later on Popper, also described the conduct peculiar to science as that which makes the losing party openly admit defeat. Were this true, then it would be admirable, as everybody agrees.

Let me mention four amusing historical examples. First, in the views of many thinkers from the seventeenth century to the twentieth, from Kepler and Galileo and Bacon to Henri Poincaré, inviting corroboration while blocking refutation is a kind of cheating unbecoming to science.[11] Second, though David Hume's theory of learning from experience says nothing about tests, much less about corroborations, in a political essay he said, since Newton's theory was tested by Frenchmen who evidently were very eager to see it refuted, their failure in this respect is most reassuring.[12] My third example is the preface of Antoine Lavoiser's classic *The Elements*, where the reader is invited to criticize his theory, as this is the best means of achieving rational assent (see Preface to Lavoisier's *The Elements* 1790). Of course, soon a valid criticism of it

---

[10] The reference is to the essay by Magee in this volume.

[11] The best presentation of the claim that one should not allow in positive evidence while blocking negative one is in Galileo's first dialogue, beginning of the fourth day. See, however, also Poincaré (*Science and Hypothesis* 1952, p. 136) and elsewhere in that book. The point hardly needs stating. It is really now generally accepted. Thus, screen actress Joan ("Alexis") Collins, hardly an intellectual innovator, said, since media-criticism is biassed against women, she pays little attention to it, and, she added as a mere after-thought, by the same token she also pays little attention to praise.

[12] "The severest scrutiny, which Newton's theory has undergone, proceed not from his own countrymen but from foreigners; and if it can overcome the obstacles, which it meets with at present in all parts of Europe, it will probably go down triumphant to the latest posterity" (Hume 1964, p. 183).

was found, but this is another matter. Finally, the latest test of general relativity. When the idea of that test was discovered, two laboratories rushed to perform it in a competition for priority. This raised the question, was the test prepared rather shoddily? The matter was decided by evidence that showed the test sufficiently accurate to serve as a possible refutation – just in accord with Popper's theory. (See for all this Pound 1979, pp. 132-139, pp. 136-137.)

This, however, is not to endorse Popper's theory of corroboration as is: it is too severe. It denies that partial corroboration can be scientific corroboration proper, contrary to the case of Newton's theory of the tides, Bohr's model of the atom, and Yukawa's theory of the meson: these were refuted yet also partially corroborated and significantly so. Eddington's famous solar eclipse observation that has impressed Popper so did not yield results that accord with Einstein but the ones in better accord with Einstein than with Newton. It was John Watkins who first noted that it is unreasonable to expect full corroboration with numerical results, contrary to the received theory of inductive support. It was John Watkins who noted that there is hardly any likelihood that a quantitative measurement will fully agree with the predicted quantity, and that this makes the inductive support of a quantitative theory practically impossible.

5. Vox Populi

This is not a general view of Popper's reputation, since this is not a beauty contest but a discussion of the corroboration as a boost to the aims of science. The disapproval of Popper's views of science exhibited by some researchers is a fact, and it raises the question, is this disapproval not a refutation? This question was raised by Pierre Duhem: since most researchers express loyalty to the inductive method, he asked, is his opposition to it not thereby refuted?[13] Many philosophers of science say, yes. The leading philosophers of science of the last half-a-century, who forge the received opinion in the field, have spread and sustained an appraisal of Popper's theory, whose aim is worldly rather than spiritual. The best example here is Moritz Schlick, who offered both a comprehensive view of the matter at hand and who pronounced an

---

[13] See Duhem (1991, chapter 4, section 8). See also the new (1990) Introduction (to the Princeton University Press 1991 edition of this book) by Jules Vuillemine, where Duhem's philosophy of science is contrasted with his own history of science (p. xxix).

appraisal of Popper's theory that is still extant. Also, Schlick was a physicist, a philosopher and an organizer of a philosophical club, the once so famous and prestigious Vienna Circle, and so he cannot be accused of ignorance of science and of its customs.[14] That he was the initiator of the traditional view of Popper that still is received and maintained makes it possible to center the present discussion around his appraisal, even though he did not express it in print; it was expressed in print by Rudolf Carnap,[15] by Wolfgang Stegmüller,[16] and by Adolf Grünbaum.[17]

---

[14] This is not to offer a comprehensive appraisal of whole of Schlick's output. His work of the period prior to his conversion to Wittgenstein's philosophy is of a distinctly different quality; this stands out when one compares Einstein's poor view of the whole neo-positivist literature with his high praise of Schlick's critique of Ernst Cassirer's works on the theory of relativity, quoted in Coffa (1991, p. 189). I always wondered how could verificationism survive Einstein's refutation of the Newtonian system that is still the paradigm of a scientific theory. It is clear, however, that I was in error: it was a last-ditch effort to save the ancient and (rightly) most respected theory of science as rationality and of rationality as verification. (The same holds for Bridgman's operationalism [1927] 1951, chapter 1, particularly pp. 3-9), where operationalism is introduced – which is therefore very different from what Mach and the young Einstein had advocated: Mach was concerned with the expulsion of metaphysics, not with absolute certainty, and Einstein was concerned with the presentation of the physical aspect of the Lorenz transformations which Lorenz had presented as an algebraic device, while presenting the formula of Newtonian kinetic energy as an approximate to the relativistic equation of mass and energy, thus abolishing certitude for good.) All this shows what a great breakthrough was Popper's idea of science as attempts at refutations and its follow-up, his theory of rationality as attempts at criticism. Nevertheless, I can scarcely find any interest in Schlick's efforts to retain the old idea, as well as those of members of his celebrated Circle, with the exception of Otto Neurath, who deviated from the official doctrine of that circle, even though (evidently for political reasons) he did not say so out loud: for, he followed the ideas not of Wittgenstein, but of Pierre Duhem, which are no doubt quite ingenious, even though they too were refuted by Einstein, as I have argued in my *Towards an Historiography of Science* (Agassi [1963] 1967, chapter 9).

[15] Carnap (1963, replies, p. 87) introductory remark to the replies to Popper:

> As Popper remarks, even at the time I thought he tended to exaggerate our differences. Today more than ever I am convinced that this is the case.

The explanation for this comes on p. 878:

> The difference between Popper's and our problem can now easily be specified . . .

The concern of Popper, Carnap explains, is with the demarcation between science and pseudo-science, whereas that of Carnap is between the empirically meaningful and the empirically meaningless. This last differentiation is possibly true: it is false if and only if empirical meaning is identified with scientific character. This identification seems indeed to be endorsed by Carnap in *Testability and Meaning* of 1936. Carnap identifies the two, indeed, there he presented Popper's view as an answer to what in his "Replies" he presents as his own problem but not as Popper's. It is clear that Carnap trims his presentation of Popper to the wind.

See also Schilpp (1959/60, pp. 490-491), where Schilpp quotes (in his own translation) Einstein's reply to him (of 19 May, 1953), in response to an invitation to contribute to *The Philosophy of Rudolf Carnap*:

> Dear Mr. Schilpp:
>
> It is a good idea to devote a volume in your series to Carnap's life-work. But I cannot accede to your request. For I have dealt with this slippery material only when my own problems have made it absolutely necessary. But even then I only studied little of [its] literature; so that I would not be able to do justice to this swarm of incessantly twittering positivistic birdies. What I have thought about it independently I have repeatedly put on paper and published (already too much!).
>
> *Entre nous*, I think that the positivist nag, which originally appeared so fresh and frisky, after the refinements which it had of necessity to undergo, has become a somewhat miserable skeleton and has become addicted to a fairly dried-up pettifoggery. In its youthful daysit nourished itself on the weaknesses of its opponents.

Now it has become respectable and finds itself in the difficult position of having to make a go of its existence under its own power, the poor thing.

[16] Stegmüller's observation is published in the homage to Carnap, republished in Hintikka (1975):

> The attempt to achieve conceptual clarity and distinctness of thought, and to broaden the spirit of rational criticism bound Carnap closely to many philosophers-friends, across lines of theoretical difference of opinion . . . If one considers Carnap's dispute with Popper in its basic philosophical terms it comes to no more than "a small internal family quarrel," which in retrospect appears much less dramatic than it appeared at the time. In their affirmative attitude to rational philosophy both of these thinkers completely at one with each other.

It has to be admitted that even Stegmüller and I are members of the same family – the family whose members favor rational philosophy; yet I shudder at the thought that this makes the differences between us "small," whatever this predicate exactly means. My review of the book in question and of the Homage there is republished in my *The Gentle Art of Philosophical Polemics* (1988).

[17] I have discussed Grünbaum's abuse of Popper in my review of his book on psychoanalysis, in my *The Gentle Art of Philosophical Polemics* (1988). But perhaps I am unfair, as he hardly refers to Popper's disagreements with his Viennese peers, and simply calls him a myth maker and one engaged in a sleight of hand. To compensate for this let me cite another case, too complex to discuss, pp. 154-155 of Gerhard Frey (1982). It is a reinterpretation of a sentence from Schlick's *Positivismus und Realismus* of 1930.

> And then follows a peculiar observation: "The verification is *logically* possible, whatever might be the case as regards its actual execution." This doubtlessly is a slip, I believe, for "verifiability (or falsifiability) of a claim is logically possible (or logically impossible)" can only mean [as a condition for Frey's assent to it] that the verifying (or falsifying) actions are in accord with (or conflict with) the theoretical and practical knowledge which is accepted at the time . . .

Since Schlick did not express his appraisal of Popper's work in print, and since my sole source for it is Popper's testimony, this may be questioned. As a matter of fact, it was questioned by John Watkins, who is more on Popper's side than on Schlick's. I therefore am willing to consider the appraisal of Popper's contribution that I ascribe to Schlick as the one originated by those who have expressed it in print, namely, Carnap, Stegmüller and Grünbaum. Yet the general view of the matter that I ascribe to Schlick are unquestionably his, and they seem to me to be unquestionably popular.[18] A critical examination of his view is of some interest to those who endorse it, and I testify that most of those with whom I discussed the matter side with him, at least much more so than with Popper. And I wish to examine the appraisal of Popper's work, be it his or that of the others.

The appraisal in question is that Popper's theory is not as unusual as he presented it, that his presentation aimed at more success than he deserves. This appraisal has the merit that it suggests that the rewards to researchers should be proportional to the significance of their

---

> The commonplace has been observed by many others, among them Popper, that all life consists in solutions to problems and tasks . . . we nevertheless do not find either in Popper or in the work of the Vienna Circle . . . they all seem to forget or ignore the fact that problem-solving . . . is not to be restricted to merely cognitive tasks . . . seems to be Schlick's view . . . The quest for absolutely certain knowledge appears herewith as a necessity of life . . .

This extraordinary and extraordinarily complex claim is simply that Popper did not discuss practical problem-solving whereas Schlick did, and even realized that "absolute" certitude must be achieved by hook or by crook or life will cease to be. This idea is as old as the response to skepticism. It is ascribed to Schlick because as he finally admitted verification to be non-existent, he had to be ascribed some weaker version of it, even at the cost of misreading a sentence in his essay, as the next note makes amply clear.

[18] A note from Alberto Coffa's book (1991, p. 421), is enlightening:

> Schlick added an un-Wittgensteinian link between verification and a personal sense of fulfillment: "The joy in knowledge is joy in verification, the exaltation of having guessed correctly . . . The question concealed behind the problem of the absolutely certain foundation of knowledge is, so to speak, that of the legitimacy of the satisfaction which verification fills us with" (Schlick 1980, p. 383). Schlick, no doubt, had in mind the average physicist making an average laboratory prediction. But some may also predict death from cancer or nuclear holocaust.

Coffa's criticism is unanswerable, of course, yet even after Coffa's two suggested qualifications of Schlick's thesis are admitted, the thesis is still unacceptable – for the reasons discussed in the body of this essay. Moreover, in the passage cited in the previous note, Gerhard Frey suggests obliquely that Schlick gave up the idea of verification proper and fell upon the idea of the absolute indispensability of certitude in practical life, including research.

contributions. The appraisal has the defect that it is clearly motivated and yet it ascribes to Popper a motive.

Since almost all participants in this unpleasant affair are dead, it is not a matter of redress any more. Also, despite frustration and aggravation from this affair, Popper had a highly successful life, measured in spiritual and worldly standards, as well as by his own standards and assessments. Yet the appraisal raised a worthwhile question: is Popper's view of any significance, and if so, what is it? On some of this matter I intend to speak now, in the hope that the current leadership will criticize me or correct their error or be removed from the helm.

6. Schlick's Passion

However controversial the history of the dispute is, and, however, live is the controversy about science and the place of corroboration in it, the general points made thus far regarding success and its criteria are quite trivial. Nevertheless, even famous thinkers have not paid them sufficient heed and said things that look quite different from these trivial points, and may contradict them. In particular, Moritz Schlick clearly held a different view. There is no greater pleasure, he said, than that associated with having made a prediction that comes true; hence, the chief role of the philosopher, he added, is cautionary: to find to prevent self-deception on this. Schlick's assertion is simply not true: many people derive much more pleasure from reading the greatest ideas than the from their own predictions, be these successful or not. Perhaps Schlick would not disagree with this factual report, but the statements he did make is, nevertheless, contrary to it, or at least it is ambiguous.

The ambiguity in Schlick's presentation is not very pressing in the context of a discussion of the appraisal of Popper's theory, however, as it is no prediction. Nor did Schlick himself or any of the members of his famous circle make any prediction. What then was Schlick's criterion for appraising them? I do not know, except that it certainly is not his verification principle, since this is a criterion for scientific character, and neither Schlick and his friends, nor Popper, had any scientific pretense, and so none of their ideas was meant to have passed it. Still, Schlick did offer a criterion of sorts, and Carnap and Stegmüller and Grünbaum seem to have agreed with him as they applied it to their assessment of Popper's theory. They said, Popper is not as different from them as he made it appear to be; and at least Grünbaum, if not Schlick, did impute to Popper a motive – that of self aggrandizement. This appraisal is very weak: there is no independent evidence as to Popper's motive, and there is no theory of the distance between theories, not to mention the ambiguity of Schlick's idea of the pleasure of right prediction. Clearly, however, the pleasure discussed by Schlick, even the pleasure one has because of one's having made a most spiritual contribution, is itself not quite spiritual, as it has to do with recognition, whereas the spiritual pleasure is impersonal: it has nothing to do with the public's opinion about its originator. It thus may be suspected that in his philosophical works Schlick himself displayed self-deception in expressing a wish for worldly success in lieu of the spiritual success that he clearly missed.

7. The Pleasure of Being Right

How is self-deception possible? How can one desire one thing and go for another? Indeed, the standard economic theory of consumers' choice precludes this possibility. Opposing this, psychoanalysis deems us as struggling seldom for what we really desire. It is hard to ascribe to researchers unconscious aims of the kind Freud ascribes to neurotics, even on his assumption that research is a substitute for sexual activity, even if researchers were generally very neurotic. Freud says we are all prone to self deception; Schlick agreed. This is not to say that Schlick's position is as comprehensible as Freud's.

At least I do not quite understand Schlick. I like to identify a piece of music played on the radio; I do not listen to the announcer saying what it is first, so that I can try to guess and listen to the announcer's repetition. When I get it right I rejoice, but I know that this is infantile, I know that I often guess wrongly or not at all, I need no police to prevent me from self-deception, and I think this is marginal in my life no matter how much I love music and enjoy the guessing game about it. Why did Schlick feel so differently? Why did he think guessing right is such a passionate business and why did he think self-deception about it is so easy and so dangerous that it needs policing? Evidently, because the scientific predictions that he was speaking of were important. What he had in mind is clearly a matter of a scientific discovery proper, whatever this is. Is the discovery so predictable? Is refutation never a discovery? Is it ever more enjoyable to be right, and is this so because it is important that a prediction is right, as Schlick suggests? Karl Popper said that it is more important that a false prediction should be refuted than that one should enjoy a prediction come true. Schlick dismissed this as masochistic, and if not he, than let me report that many others die Schlick did speak of the joy of a true prediction, but clearly he did so on the basis of the hypothesis that it is corroborated prediction that proved to be scientifically important. Is this hypothesis corroborated?

Though the comment on Popper's theory that I ascribe to Schlick and the response to it that I ascribe to Popper were not previously published, as far as I know, I deem them very important even for those who deny my ascriptions. In addition to Popper's response to Schlick there is his view on the personal side of the matter: he suggested that if my prediction is false, I prefer that I refute it than that others do so. It is hard to decide how serious this suggestion was with Popper, as he said so only in his seminars and in private sessions. Nevertheless, let me report, in my very last and very lengthy conversation with him about our differences of opinion at large he stressed that he found my view on corroboration unacceptable just because it excludes the demand that he found of supreme importance, that one should feel strongly the desire to avoid error. Unlike Bacon, he rejected the corollary from this that it is desirable to avoid voicing a hypothesis. Hence, he had to find it desirable to be corroborated. Unlike Bacon, he found it possible that error be corrected. And he added in that conversation that it is desirable, to wish to correct one's errors as soon as possible. He concluded then that it is preferable that one should feel unhappy about one's errors. I strongly dissented from this conclusion then, and I still do. And if I place too much stress on points made by Popper in seminars and in private sessions, then let me say that it is easier forme to say of my own view that for my part, such as it is, I barren to consider it important to distinguish between regrettable error concerning which one should, indeed, be unhappy, and understandable error, of which one can and should be proud. And I even consider some deep criticism causes for personal joy, and I am in deep debt to Karl Popper that he made it possible for me to advance this view while taking account of the importance of scientific progress and of the role of corroboration in that process.

For the sake of the record, let me add that Popper himself was in two minds. Alan Musgrave reports[19] that in his seminar Popper expressed

---

[19] Professor Musgrave notes:

I did not report Popper's pride in being refuted.
I reported thus: [Pavel] Tichy concluded [his seminar talk] "Popper's definition of verisimilitude is worthless." Popper said "I disagree with only one word – your last. Nothing is worthless which has provoked such an excellent criticism."

Alan Musgrave.
23 November 1994
I am grateful for his permission to reproduce this note.

pride in having developed a theory that was beautifully refuted. I report
the opposite. The two reports pertain to the same refutation, since only
two of ideas of Popper's large output were unquestionably refuted. But
let us leave history for now.

8. The Pain of Being Criticized

I often say in public that some criticisms of one's views should make one
rejoice, and I often add the question, why is this not taken for granted as
it should be? Usually my audiences are astonished to hear this. They say
that I must know the answer, that I only tease them when I pretend
otherwise. They say I must be aware of the fact that it is silly of me to
compare recognizing a musical composition or style with a new scientific
hypothesis, that I compare criticism made in private to that made public,
that I ignore the fact that everyone can guess incorrectly yet guessing
correctly is at times a cause for a Nobel Prize. They then look at me
triumphantly and eagerly await my admission of guilt.

This is funny. People who respond this way do not even notice that
this way they speak of worldly success, and they would disregard as
irrelevant to their response to me any adverse comment that I and others
may make of the Nobel Committee. Convinced that they know the
answer, and giving a simple explanation for it, they show by giving this
very explanation that they do not understand it at all, that they overlook
the fact that the desire for the Nobel Prize is a motive different from the
desire to be right. To see this most graphically, all one needs is to devise
a case of choice between the two: suppose one has the choice between
corroborating one's conjecture and gaining the Nobel Prize. What would
one choose then?

This argument is so forceful that the natural response to it is to reject
it as too fantastic: as the Nobel Prize is won for being right, not for being
wrong, it is impossible to construe, even most hypothetically, that one
can have the choice between proving oneself right and winning the Prize.
Yet this often is the case. One such case should suffice, the refutation of
Pauli's theory that the elements that are noble (i.e., they cannot combine)
are those whose outer shells are full. Indeed, it seems to me obvious that
the novelty of a fact is guaranteed by its conflict with a received
scientific theory.

And yet, there is no escape from the fact that the response which this
evidence contradicts sounds most convincing. The Nobel Prize is
awarded to the originator of conjectures, only if they are good, namely,

corroborated. This last statement concerns both the Nobel Committee and the goodness of theories. Presumably the members of that committee require corroboration before agreeing on an award. But are all good conjectures corroborated and are all corroborated conjectures good? This depends on the aim of science. If it is corroboration, then, clearly, yes. If it is any other aim, does this aim requires corroboration as means? If yes, then, again, yes. Otherwise, perhaps not. The paradigm of a conjecture recognized as important even though it was never corroborated is the famous Bohr-Kramers-Slater view of the law of conservation of energy as restricted to statistics: it was refuted on its very first test. Why then was Einstein's theory of gravity appreciated only after Eddington's observation corroborated it? Because most researchers think that to be good the hypothesis requires a corroboration. Nobel Prize winner Max Born wrote, and in public lectures stressed, that researchers first guess and then try to confirm their guesses. I do not know how seriously he said this: Schrödinger dismissed this saying, he would not have repeated it under oath. Yet he did write and say this and it impressed me enough to remember it and wonder about it for many years. I also remember how many researchers commented on wrong guesses with dismay, as if their job was to guess not just intelligently, but also rightly, to win a gamble. But what is a win? Let me repeat: ever so many important wins, allegedly right predictions, are full of errors. And ever so many refutations would not have been made but for the theories they refuted.

This discussion should be truncated, since it involves both the matter of the psychology of criticism and the nature of correct prediction in science. The simplest way to do this is to observe that the following simple thought experiment fully accords with the details marshalled here, including the fluff that was Schlick's philosophy: public criticism would be greatly coveted for worldly reasons, were it the rule of the Royal Society that people openly criticized by Fellows of that Society should be invited to join it, or were it the rule that the public criticism by a Nobel Laureate would automatically place its target on the list of candidates for that coveted prize.

Proof: most researchers guess that the Nobel prize is out of their reach; they are all happy to be refuted on this matter.

So what did Schlick mean exactly?

Presumably Schlick was not speaking of the psychology of finding any of one's guesses true. If one guesses that one is dying, that one's friend has betrayed one, then one is presumably happy to be refuted. Clearly, he was speaking as a person whose life was so sedate that the only thing that could improve his lot was scientific success. And, to repeat, he identified this with corroboration. It is time to explode, then, the equation of scientific success with corroboration.

Let us then leave desire in favor of desirability. What is the desirability of corroboration in science? How is its desirability to be understood?

9. Corroboration as a Scientific Success and as a Scientific Failure

The case of a prediction of a disaster can be further elaborated on. Both Jonah and Cassandra foresaw disaster. The source of their foresight was not science but divine revelation, but let us ignore this for a while. Cassandra took it for granted that the disaster could be prevented and she therefore made a public prediction. Yet as divine revelation is true, her effort had to be in vain: she was not believed and so no one tried to avert the disaster. Not so Jonah. He foresaw a disaster and preferred not to predict it to the people in question. Indeed he ran away and would have died but for being swallowed by a big fish and thus was given a second chance. He did make the prediction, was believed, caused repentance and thus caused the prevention of the disaster. He was peeved and the Lord chided him for having taken the public refutation of his own prediction as more aggravating than the disaster that he had thus averted. The story of Jonah makes it clear that the psychology of prediction is different from the public aspect of it, and that though one should be delighted in the refutation of an ill prediction, as a refutation it is a cause for grievance, even though the grievance is small in comparison with the success of averting a disaster.

The explanation for Jonah's grievance is methodological. Martin Buber stressed that the moral from this story is that prophets are not meant to be right but to preach repentance. Popper disagreed. He distinguished between prediction and prophecy: prediction is a test of a theory from which it follows; refuting a prediction, therefore, refutes a theory. Not so with prophecy: prophecy come true is the evidence of the potency of the God who stands behind it, and prophecy refuted refutes no theory and so is unenlightening. The preference for a prophecy come true is thus replaced by the preference for critical dialogue.

This may be true, but it is insufficient: even if critical dialogue is preferable, it is also possible to deem preferable a dialogue with no defeat to one with defeat. This sounds convincing as it is true with one hidden proviso: it is better to hold a true theory than a false one, and then hopefully one will not be defeated in a debate. It is only because we are fallible, says Socrates in Plato's *Gorgias*, that it is good for one's soul to be refuted. Why, then, do we have the intuition that even if our conjectures are false that it is better to have them corroborated?

This is, of course, the major problem of this discussion, and we bump into it here repeatedly. A few partial answers were offered here: intuition misleads, corroboration is erroneously taken to be evident that one's conjecture is true, worldly success is offered to the one whose conjecture is corroborated. Yet this is not the whole story. The truth of the matter is that only very unusual hypotheses invite empirical criticism, and these are very successful even before they are tested. There is no conjecture that invited new tests that was not historically valuable regardless of the outcome of its test. Moreover, it is *a priori* hardly conceivable that a successful theory will not be corroborated one way or another if we allow partial corroboration such as the one allotted to Newton's theory of the tides, Bohr's model of the atom, and Yukawa's theory of the meson. Of course, the corroboration of Einstein's theory of gravity made it ever more powerful competitor to Newton's, as the defenders of Newton's theory had quite a task on their hand. This, incidentally, is why though the corroboration of Einstein's theory was only partial it counted for so much.

Schlick's error, then, was in his lack of a sense of proportion, not in the idea that corroboration is pleasing: at times it may indeed be pleasing, and then it should please ones whose views are refuted too, though understandably not so much as their competitors. Indeed, I should observe that even in Popper's writings he always refers to the corroboration of his ideas as pleasing. Nevertheless, for my part, this is not to approve of taking it as generally correct; this I find hard to swallow. Is it too much to expect a researcher rejoice so much in progress without paying much attention to the fact that they were refuted? At least Kepler wrote to Galileo that the latter's discovery of new moons only delighted him and led him to new conjectures. These too were refuted, incidentally.

Had Kepler returned to life and seen how far his ideas receded into the background, would he be sad? I do not know. Faraday said, even though such experience would not make him happy, he would welcome it. But why would it make him unhappy? Many students and colleagues tell me that this is in human nature. At least Popper and Grünbaum would agree that we should not have dogmatic opinions on what is human nature but make it a matter for empirical investigation. As Professor Suárez-Iñiguez said, much of what was taken to be matters of human nature turns out to be matters of education, given to empirical tests. The received view of scientific support is admittedly deeply connected with the view of human nature as defensive, much as described by Sir Francis Bacon: the wish to be proven right he repeatedly declared, is what makes us defensive. Schlick's idea is, I presume, not far from Bacon's, even though less felicitously expressed, since Bacon concluded that we should speak cautiously, yet Schlick demanded that we take risks but succeed. The proper critical attitude is to disfavor defensiveness by commending the courage of the proponents of important refuted ideas. If so, then the diffusion of Popper's view of corroboration should bring about far-reaching reforms. It promises the rise of a new era.

Tel-Aviv University
Philosophy Department
P.O. Box 39040
Ramat Aviv, Tel-Aviv 69978,
Israel
e-mail: agass@post.tau.ac.il

REFERENCES

Agassi, J. ([1963] 1967). *Towards an Historiography of Science, History and Theory.* Beiheft 2.
Agassi, J. (1971). *Faraday as a Natural Philosopher.* Chicago: Chicago University Press.
Agassi, J. (1975). *Science in Flux, Boston Studies.* Dordrecht: Kluwer.
Agassi, J. and I.C. Jarvie (1987). *Rationality: The Critical View.* Dordrecht: Kluwer.
Agassi, J. (1988). *The Gentle Art of Philosophical Polemics.* La Salle, IL: Open Court.
Bernard, C. (1940). *An Introduction to the Study of Experimental Medicine.* Ann Arbor, MI: Edwards Brothers Inc.
Bridgman, P.W. ([1927] 1951). *The Logic of Modern Physics.* New York: Macmillan.
Budworth, D.W. (1981). *Public Science Private View.* Bristol: A. Hilger.
Carnap, R. (1963). *The Philosophy of Rudolf Carnap.* Edited by: Paul Arthur Schlipp. La Salle, IL: Open Court.
Coffa, A.J. (1991). *The Semantic Tradition from Kant to Carnap to the Vienna Station.* Cambridge: Cambridge University Press.
Duhem, P. (1991). *The Aim and Structure of Physical Theory.* Translated by Philip P. Wiener. Princeton: Princeton University Press.
Faraday, M. (1839). *Experimental Researches in Electricity.* London: Bernard Quaritch.
Frey, G. (1982). Schlick's Konstituirungen as a Special Case of General Theoretical Principles of Scientific Inquiry. In: Eugene Gadol (ed.), *Rationality and Science: Memorial Volume for Moritz Schlick in Celebration of the Centennial of his Birth,* pp. 145-154. New York: Springer.
Hintikka, J., ed. (1975). *Rudolf Carnap Logical Empiricist: Materials and Perspectives.* Dordrecht: Kluwer.

Hume, D. (1964). On the Rise and Progress of the Arts and Sciences. In: *Essays, Moral, Political and Literary*, vol 3, p. 183. Aalen: Scientia Verlag.

Jarvie, I.C. (1982). *In Pursuit of Truth: Essays in Honor of Karl Popper's 80th Birthday*. Edited by: Paul Levinson. Atlantic Heights, NJ: Humanities.

Lavoisier, A. (1790). *The Elements*. Translation by Robert Kerr. Edinburgh: W. Creech. Reprint: New York: Dover Publications, 1965.

Poincaré, H. (1952). *Science and Hypothesis*. New York: Dover.

Pound, R.V. (1979). Terrestrial Measurements of the Gravitational and Red Shift. In: Gerald E. Tauber (ed.), *Albert Einstein's Theory of General Relativity*, pp. 132-139. New York: Crown Publishing Co.

Popper, K.R. (1957). *The Poverty of Historicism*. London: Routledge.

Popper, K.R. (1959). *The Logic of Scientific Discovery*. London: Hutchinson.

Romo, J. and Doncel, M.G. (1994). Faraday's Initial Mistake Concerning the Direction of Induced Currents, and the Manuscript of Series 1 of his Researches. *Archive for History of Exact Sciences* **47**, 291-386.

Schilpp, P.A. (1959/60). The Abdication of Philosophy. *Kant-Studien* **52**, 480-495.

Schlick, M. (1932/33). *Positivismus und Realismus. Erkenntnis* **3**, 1-31.

Schlick, M. (1980). *Philosophical Papers*, vol. 2. Edited by H.L. Mulder and B.F.B. van de Velde-Schlick. Dordrecht: Kluwer.

John Watkins

POPPER AND DARWINISM

I

The first Darwin Lecture was given in 1977 by Karl Popper. He there said that he had known Darwin's face and name "for as long as I can remember" (1978, p. 339); for his father's library contained a portrait of Darwin and translations of most of Darwin's works (1974, p. 6). But it was not until Popper was in his late fifties that Darwin begin to figure importantly in his writings, and he was nearly seventy when he adopted from Donald Campbell the term "evolutionary epistemology" as a name for his theory of the growth of knowledge (1972, p. 67). There were people who saw evolutionary epistemology as a major new turn in Popper's philosophy.[1] I do not share that view. On the other hand, there is a piece from this evolutionist period which I regard as a real nugget.

I call it The Spearhead Model of evolutionary development. It appeared briefly in the Herbert Spencer lecture he gave in 1961, which he wrote in a hurry and left in a rough and unready state. It contained mistakes that would, and did, dismay professional evolutionists. Peter Medawar advised him not to publish it,[2] and it lay around unpublished for over a decade. He eventually published it, with additions but otherwise unrevised, in Chapter 7 of *Objective Knowledge* (1972). It did not, so far as I know, evoke any public comment from biologists or evolutionists. I touched on it briefly in my contribution to the Schilpp volume, but I know of no other published discussion of it.

When I discussed the neglect of Popper's Herbert Spencer lecture with Bill Bartley, in 1978, he was pretty dismissive, saying that it was all in Alister Hardy. Now it is true that much of that lecture is about an idea,

---

[1] I am thinking especially of the late Bill Bartley; see Radnitzky & Bartley (1987), part 1.

[2] He is the expert mentioned by Popper on p. 281 of his (1972).

In: E. Suárez-Iñiguez (ed.), *The Power of Argumentation* (*Poznań Studies in the Philosophy of the Sciences and the Humanities*, vol. 93), pp. 101-116. Amsterdam/New York, NY: Rodopi, 2007.

which Popper was later (1982, pp. 39*f*) to call "active Darwinism," which had indeed been anticipated by Hardy. Hardy had suggested that an animal's interests might change in such a way that certain bodily mutations that would previously have been unfavourable now become favourable; and he added that while such a change of interest might be forced upon the animal by external circumstances, it might result from exploratory curiosity and the discovery of new ways of life. He gave the example of forebears of the modern woodpecker switching their attention from insects in the open to insects in the bark of trees. No doubt these proto-woodpeckers proceeded rather clumsily at first; but going after this rich new food supply with tools not well adapted to the task proved at least marginally more rewarding than persisting with old habits. So new habits developed; and mutations that made the bird's bodily structure better adapted to these new functions now became advantageous, though they would previously have been disadvantageous. So new shapes of claws, beak, tongue, etc., began to evolve. Popper put forward the same idea, together with the woodpecker example, without reference to Hardy. But I am sure that Popper was entirely innocent here. He gave this lecture in 1961. Hardy presented the idea in his book *The Living Stream* (1965). Before that he had broached it in the Linnean Society's Proceedings in 1957, but Popper would not have seen that. If Popper made a mistake it was to publish this lecture essentially unrevised in 1972. When he sent Hardy a copy of *Objective Knowledge* he apologized for its lack of any reference to him.[3] He made ample amends later in his (1975), (1977), and (1982).

In any case the Spearhead Model is distinct from "active Darwinism." It is about certain relations between an animal's central control system and its motor system. Popper was usually a good publicist for his own ideas, and he subsequently gave plenty of publicity to the idea of "active Darwinism." But he allowed the Spearhead Model to fall into neglect. I will resurrect it. In case you are wondering whether a neglected contribution to evolutionary theory is a suitable subject for a philosophical lecture, I may add that the Spearhead Model has important implications, which I will try to spell out, for the mind-body problem.

But first I will say a few words about the relation between Darwinism and Popper's theory of knowledge. Against my contention that Darwin's ideas do not seem to have had a serious impact on Popper before about 1960 it might be objected that their influence had shown itself already in *Logik der Forschung* (1934); for Popper there declared that the aim of the

---

[3] I am here drawing on the Popper papers.

method of science is to select among competing hypotheses "the one which is by comparison the fittest, by exposing them all to the fiercest struggle for survival" (1959, p. 42). And I agree that there is a partial analogy between his conception of scientific progress through conjectures and refutations and Darwin's conception of evolution through variation and natural selection. But there are also serious disanalogies. Perhaps the main one is this. According to Darwin, any *large* variation is sure to be unfavourable; to have any chance of being favourable, a variation has to be very slight. And this of course means that evolutionary developments are gradual and slow. But science during the last four centuries has been evolving, if that's the word, by leaps and bounds. *Inductivism* may view scientific progress as a smooth, incremental process, but Popper's view of it is essentially saltationist, the new scientific theory usually conflicting radically with its predecessors at the theoretical level and making small changes at the empirical level. For Darwin there could be no such thing as a "hopeful monster"; but the history of science, seen through Popperian eyes, is full of "hopeful monsters." As I see it, the relation of a revolutionary new scientific theory to its predecessors is not unlike that of a jet-engined aircraft to its propeller-driven predecessors; the new theory takes over the work done by them, and does it better, being driven by a more powerful theoretical ontology. But as Richard Dawkins pointed out (1982, pp. 38-39), for a jet-engined aircraft to have evolved from a propeller-driven one in a *Darwinian* manner, each nut, rivet and other small component of the earlier plane would have had to change, one at a time and by series of very small steps, into a component of the later one.

I do not find Popper paying much heed to Darwinism in his middle years. There are no significant references to Darwin or Darwinism in *The Open Society* (1945) and no references in *Conjectures and Refutations* (1963). There is a brief discussion of Darwinism in "The Poverty of Historicism" (1944-1945), but its tendency is rather deflationary. Understandably anxious to dispel the idea that evolutionary theory gives any support to the historicist thesis that society is subject to a law of evolution, he said that it has "the character of a particular (singular or specific) historical statement. (It is of the same status as the historical statement: 'Charles Darwin and Francis Galton had a common grandfather'.)" And he endorsed (1957, p. 106n) Canon Raven's dismissal of the clash between Darwinism and Christianity as "a storm in a Victorian tea-cup" (Popper 1957, pp. 106-107*ff*).

But by the time he was writing his intellectual autobiography for the Schilpp volume, around 1968-69, the situation had changed. He devoted a whole section to Darwinism, whose great importance he now proclaimed. The suggestion that it has a merely historical character was silently abandoned; but Popper vacillated considerably over what character it does have. He said that Darwinism is "almost tautological" (1974, p. 134), which doesn't sound too good, coming from someone who extols a high falsifiable content in scientific theories. He also said that Darwinism is a metaphysical research programme. Does that mean that it makes no predictions? Well, he conceded that it predicts the *gradualness* of all evolutionary developments, but he added that this is its *only* prediction. But there he was surely wrong. If a species gets geographically split in two, for instance by geological changes, and there are considerable ecological differences between the two geographical areas, Darwin's theory predicts that, provided neither population becomes extinct, there will in due course be two species between which there can be no interbreeding.

Having declared it to be a metaphysical research programme, Popper added that Darwinism is not "just one metaphysical research programme among others" (p. 135). Then what singles it out? In answering this Popper went off on a new tack; and what he now said made it sound as though it was not so much that his theory had benefited from Darwinism, but that Darwinism could now benefit from his theory or at least that the benefit had been mutual. He wrote: "I also regard Darwinism as an application of what I call 'situational logic'." And he added: "Should the view of Darwinian theory as situational logic be acceptable, then we could explain the strange similarity between my theory of the growth of knowledge and Darwinism: both would be cases of situational logic" (p. 135).

I find this baffling. What he had called "situational logic" involved an agent in a well-defined situation, for instance a buyer in a market, where the agent's situational appraisal and preferences jointly prescribe a definite course of action. What has that to do with the theory of evolution? There is no inconsistency in supposing both that all creatures always behave in accordance with the logic of their situation, *and* that all species are descended unchanged from original prototypes. Situational logic has nothing to say about the two assumptions that differentiates Darwin's theory from contemporary alternatives to it, namely that heritable variations occur and that a successful variation may get preserved.

Nor does situational logic say anything about an assumption that differentiates Popper's theory of the growth of scientific knowledge from Humean and other empiricist views, namely that science essentially involves *intellectual innovation*. Indeed situational logic and intellectual innovation are inimical to one another. In October 1948 Bertrand Russell found the flying-boat in which he had landed at Trondheim starting to sink. When he got outside, still clutching his attaché case, he found that there was a rescue boat about 20 yards away which, for safety reasons, would not come any closer. He did not attempt to bring new ideas to bear on his problem, but acted in accordance with the logic of his situation, which required that he throw away his attaché case and swim those 20 yards. But when, years earlier, he had been struggling innovatively with the paradoxes he had discovered at the foundations of mathematics, he had no situational logic to prescribe a course for him; nor did situational logic provide a compass for Newton when he was "voyaging through strange seas of thought alone."

I don't think that Popper ever came up with a satisfactory answer to the question, "Why is Darwinism important?".

II

I turn now to the nugget which I discern in Popper's later, evolutionist period. Although it needs revamping, it is potentially a significant contribution to Darwinian Theory, enhancing the latter's problem-solving power. And I will begin by indicating the problem which the Spearhead Model helps to solve.

We have already met Darwin's insistence that only very slight variations have any chance of being favourable. Both T.H. Huxley and A.R. Wallace objected to this. Chapter 6 of Wallace's *Darwinism* (1889) is entitled "Difficulties and Objections"; and its first sub-heading is: "Difficulty as to smallness of variations." He there said that Darwin had exaggerated how slight a favourable variation has to be, thereby inviting the objection "that such small and slight variations could be of no real use" (p. 127). Was Wallace right? R.A. Fisher calculated that "a mutation conferring an advantage of 1 per cent in survival has itself a chance of about 1 in 50 of establishing itself and sweeping over the entire species" (1930, pp. 77-78). But a *very* slight variation would be more likely to confer an advantage of the order of 0.001% or perhaps 0.0000001%; what chance would it have of establishing itself? Darwin's gradualism seems to create the difficulty that variations will be either too large to be favourable or too slight to catch on. The Spearhead Model can be of assistance here, as we will see.

The Spearhead Model involves a dualism between central control system and controlled motor system. Think of an animal facing an urgent survival problem – a leopardess, for instance, with hungry cubs to feed. She sets out and in due course spots an impala; does her retinal image of this desirable object stimulate this powerful motor system of limbs, claws, teeth, etc. to go bounding after it? Certainly not; the motor system is under skilful and precise central control. She moves silently, keeping upwind and out of sight. When she eventually springs, she anticipates the impala's escaping movement. Although hungry, she does not at this stage feed off the carcass, but drags it back to the lair where her cubs are waiting.

According to the Spearhead Model, motor systems and control systems are genetically independent of one another. Of course, it might happen that a mutation that brings about a change in one also brings about a change in the other, but it would be an incredible coincidence if it brought about coordinated changes. And the model's main message is that in the evolution of species, *developments in control systems pave the way*.

The main objection to the claim that a large variation may be favourable is, I suppose, that the organ in which it occurs will thereby become maladjusted with respect to other organs with which it had been in equilibrium. More specifically, an organ whose size or power is enhanced by a large variation would make overtaxing demands on the rest of the system; a considerable lengthening of a wing, say, might overtax the wing muscles, and a considerable strengthening of wing muscle might overtax the heart. I now introduce a point which Popper did not make explicit. It is that this objection falls away when we turn from the animal's motor system to its control system. Considerable enhancements *there*, I assume, do not have an overtaxing effect elsewhere. Giving a computer more powerful software, even improving the hardware itself, say by adding a new chip, puts no extra demand on the electricity supply. And I will assume that much the same holds for animals' control systems. Carrying around an over-sized wing would be costly, but carrying around even a lot of surplus control capacity may be virtually cost-free.

Popper presented a thought-experiment which begins by envisaging a state, call it state (i), in which an animal's control system and motor system are, in his words, "in exact balance" (1972, p. 278); by this he seems to have meant that the control system can cope, but only just, with the motor system. Let C and M denote respectively the control capacity and the motor power of the two systems when they are in this exact balance. Popper next envisaged two alternative developments: into a state (ii) in which control capacity C remains constant and motor power M receives an increment, and into a state (iii) in which M remains constant and C receives an increment. In the diagram the thin lines stand for control capacity and the thick ones for motor power:

C	C	$C + \Delta C$
M	$M + \Delta M$	M
(i)	(ii)	(iii)

In (ii), where motor power exceeds control capacity, the result, Popper declared, would be *lethal*. What about (iii), where control capacity exceeds motor power? As Popper saw it, there would be no immediate advantage; but it might be *extremely favourable* (p. 278) later, if motor power were suitably augmented.

I will refer to this as the Spearhead Model Mk I. As it stands, it is exposed to several objections. Perhaps the main one is this. We are here interested only in very slight variations, at least on the motor side, on the Darwinian assumption that only these have any chance of being favourable. There can of course be no objection to the idea that the same slight variation may be unfavourable in one context and favourable in another; but that it should prove *lethal* in one context and *extremely favourable* in another seems farfetched. Here is a simple thought-experiment: imagine that control capacity, having come to exceed motor power quite considerably, now remains constant for a period during which there is a sequence of 100 small increments of motor power, the 99th increment bringing the system back into a new state (i) of "exact balance," and the 100th tipping it over into a new state (ii). On Popper's account all these small increments will be favourable except for the last, which will be lethal. Continuity considerations argue against such an abrupt and calamitous reversal.

The source of the trouble, I believe, was Popper's implicit use of an all-or-nothing notion of control, whereby the motor system is either fully under control or else, the moment its power rises above a critical level, quite out of control. That is what generates the abrupt switch from favourable to lethal; and it has the unwanted implication that, given a motor system that is fully under control, no enhancement of control capacity will bring any advantage while the motor power remains constant. Then why should mutations get preserved which merely endow the system with potential control over motor power it doesn't yet have? Popper side-stepped this question. One of his sentences began: "Now once a mutation like this is established . . . " (p. 278), but *how* it might get established he did not say. He allowed that mutations of this kind are "only indirectly favourable," but claimed, as we saw, that once (somehow) established they may prove *extremely* favourable. But natural selection lacks foresight; a gene that is not now favourable will not now be selected for, however favourable it would prove later if it were selected now. One begins to see why evolutionists did not find this model persuasive.

The suggestion that a motor system abruptly switches from being fully under control to being quite out of control the moment its power rises above a critical level presupposes that its power is always used to the full. But surely power can be exploited judiciously. When her Mini won't start, a housewife borrows her husband's Jaguar to go shopping. This car would soon be out of control if a driver kept its accelerator fully depressed; but she is content to draw on only a fraction of its reserves of power. However, there is a countervailing consideration here. Imagine that this Jaguar is now being used as a getaway car with a police-car in hot pursuit; its driver may use its large reserves of power too freely, perhaps with lethal results. (Within a few weeks of first putting that thought on paper I read in the newspaper of five or six cases where a criminal was either killed or injured when his getaway car crashed.) While properly sceptical of the idea that the system will get hopelessly out of control the moment its power rises the least bit above a critical level, we can agree that the risk of its getting out of control increases the more its power exceeds its control capacity.

In an attempt to revise this model so that it complies with continuity requirements, I will now present a thought-experiment which relates variable motor power to variable control capacity. It involves racing cars of varying power, and drivers of varying skill. Imagine a fleet of 100 racing cars whose engine power ranges from enormously powerful right down to that of, say, a minicab. The cars are otherwise as similar to one another as is consistent with this very varying motor power. To drive them there is a corps of 100 drivers whose racing skills likewise range from that of a reigning Grand Prix champion down to that of, say, a minicab driver. The whole show is run by a central management, which organizes tournaments in the course of which each driver races in every car. The route, which is varied from race to race, is picked out from a complicated road network, and always includes some quite long straight stretches that end abruptly in tight turns. In the interest of safety the physical road is wide, but it has a rather narrow centre lane, painted white, on either side of which are grey areas, with black areas beyond them. Cars are supposed to stay in the centre lane. Straying into a grey zone, the equivalent of getting partially out of control, incurs a penalty stiff enough to forfeit two or three places, while straying into a black zone, the equivalent of getting right out of control, incurs disqualification. The cars are clocked as they go round the circuit, one at a time, and any straying into grey or black zones is automatically recorded. All drivers have a strong incentive to perform well, an ideal performance being to complete the circuit in the shortest time with no penalties. I assume that, as well as a metric for the cars' motor powers or M-values, there is a metric, analogous to the Elo-rating for chess masters, for the drivers' racing skills or C-ratings. The combination of a certain C-rating with a certain M-value will determine a level of *racing fitness*, as I will call it in analogy with the notion of biological fitness.

Let us now pick on a driver with a C-rating in the middle range, and ask whether for him there is a car with an optimum M-value, that is, one which maximizes his racing fitness. There will surely be cars whose M-values are too low for him. What about cars at the top end of the range? Would he do best in the car with the highest M-value of all? I say that he would not. Consider what would be likely to happen when he gets to one of those straight stretches. Being strongly motivated to win, he will want to go flat-out; but unless he judges it exactly right he will be liable to get into difficulties at the next bend. In short, there is a danger that he will resemble the desperate driver of a getaway car. Now if there are M-values that are too high for him, and ones that are too low, the principle of continuity tells us that somewhere in between there is a value that is neither too low nor too high but *optimum*.

In the diagram, variable control skill C is represented along the vertical axis and variable motor power M along the horizontal axis. Each isobar depicts the common level of racing fitness yielded by various mixes of C- and M-values. If one of these has a lower C-value and a higher M-value than another, the former's tendency to complete the circuit more quickly will be nicely balanced by its tendency to incur more penalties. (The numbers attached to the fitness-isobars have only an ordinal significance.) The line OA, which need not have been straight, represents pairings of given C-values each with its optimum M-value. (For a given M-value there is no optimum C-value: an increase of C is never disadvantageous and nearly always advantageous. The reason it is not always advantageous is that for low *M*-values further increases of *C* will eventually become futile; that is why the isobars eventually become vertical on the left.) Control capacity is running ahead of motor power in combinations above the *OA* line, and lagging behind in combinations below it.

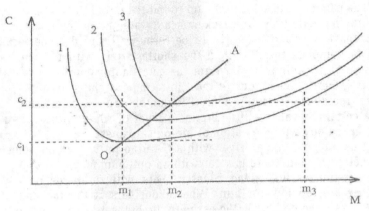

Fig. 1 The Spearhead Model Mk II

I call this the Spearhead Model Mk II. Let us now see how it overcomes the difficulties that attended the Mk I model. To the Mk I idea of an initial states (i) in which a *C*-value is paired with its best *M*-value, namely the one with which it is in "exact balance," there corresponds here a position on the *OA* line where a *C*-value is paired with its optimum *M*-value. To the Mk I idea of a shift to a state (ii), or a state (iii), there corresponds here a sideways, or a vertical, move away from a position on the *OA* line. It is with these one-sided movements that differences between the models open up. In the Mk I model, a unilateral increase in *M*, even quite a small one, drove fitness right down, while a unilateral increase in *C*, even quite a large one, brought no direct benefit. In the Mk II model, unilateral increases of *M* away from a position on the *OA*-line always reduce fitness, though in a continuous, non-abrupt manner, while unilateral increases of *C* normally bring *actual* as well as potential benefit. Thus an increase of *C* from c_1 to c_2 with *M* remaining constant at m_1 raises fitness from level-l to level-2; and this increase in *C* is also potentially beneficial in that a subsequent increase of *M* from m_1 to m_2

with C now remaining constant would raise fitness from level-2 to level-3. But if M then went on increasing unilaterally to m_3, fitness would sink back to level-1.

That an enhancement of control capacity almost always has actual as well as potential utility is the crucial difference of the Mk II from the Mk I model. If we look for real cases where an increase of control capacity has *only* a potential utility, I think that they will be found to presuppose the possibility of saltations on the motor side. Years ago I had a typewriter with a black/red control lever which was of no actual utility because the ribbon was all-black. More recently I had a computer with a colour control which was of no actual utility because the display unit was black-and-white. To become advantageous these controls called for a saltation to a black/red ribbon, or to a colour display unit. Saltations are of course foreign to evolutionary developments as understood by Darwin. The continuities which characterize these should ensure that any enhancement of control capacity that improves it vis-à-vis increased motor power in the future will also improve it vis-à-vis existing motor power: no potential advantage, we might say, without actual advantage.

Of course, this Mk II Model, with its assumption that developments on the motor side take place along only one dimension, is still terribly over-simplified. It is as if we attended only to our leopardess's fleetness, ignoring her claws, teeth, etc. But my hope is that elaborating it to reflect the multi-dimensionality of real biological motor systems would leave unchanged its central message, which is that *advances in central control systems lead the way in evolutionary developments.* This now receives a straightforward Darwinian endorsement. Unilateral increases in C will nearly always be selected for since they are never unfavourable and nearly always *actually* as well as potentially favourable. By contrast, a unilateral increase in M will be favourable only if the existing combination of C and M is above the OA line. An evolutionary development would be selected against if it strayed into the area below the OA line, but there is nothing to stop it rising way above that line. Control capacity may run far ahead of motor power, but motor power may develop only along paths marked out by its senior partner. And we can say to Wallace that slight variations *could* be of real use: a slight increase in motor power that fits in with pre-existing control capacity is likely to be exploited skilfully and intensively. Start with a number of equally skilled Grand Prix drivers in identical racing machines, and now give one car a very slight improvement to its acceleration; its driver will exploit this to the utmost, and in a close race it might just edge him into first place.

III

As a way of bringing out the philosophical significance of the Spearhead Model I will turn to the book co-authored by Karl Popper and John Eccles, *The Self and Its Brain* (1977). There are large similarities but also important differences between the two authors' positions. Both men upheld a brand of Cartesian dualism and interactionism; they both maintained that there is a ghost in the machine, and that the ghost significantly influences the machine. (I too believe this.) But whereas Popper held an evolutionist version of Cartesian interactionism, Eccles reverted to something very close to the classical dualism of Descartes himself, whereby an immaterial soul is sent from without into a naturally generated body. Eccles said that while his body is a product of natural selection, his soul or self-conscious mind has a supernatural origin (Popper and Eccles 1977, pp. 559-560). My thesis will be that the Descartes-Eccles position is afflicted by a frightful difficulty which is dissolved by the Spearhead Model, at least in its Mk II version. Unfortunately there is no mention of the Spearhead Model in *The Self and Its Brain*; it seems that Popper had already lost confidence in it by then. The fundamental disagreement between him and Eccles briefly surfaced near the end of their book (pp. 559*f*) and was then allowed to fizzle out. I will re-open it and bring the Spearhead Model to bear on it.

I will now explore the Descartes-Eccles position from within for internal difficulties, engaging only in immanent criticism and not questioning its main assumptions. Descartes had envisaged the soul playing upon the pineal gland whose movements control the flow of animal spirits and thence muscular contractions and bodily movements. He claimed that in human beings the pineal gland, being *very* small, is easily moved. In Eccles's answer to the question: "How does my soul come to be in liaison with my brain that has an evolutionary origin?" (p. 560), what he called the "liaison brain" plays a role quite analogous to that played by the pineal gland in Descartes's system. The main difference is that for Descartes the pineal gland has a fixed location within the brain, whereas for Eccles this liaison role is played by different parts of the brain at different times. Eccles claimed that the liaison brain, being at any given time only a minute fraction of the whole brain, is susceptible to the "weak actions" by means of which the mind intervenes upon the neural machinery to bring about voluntary actions (pp. 356-364).

The great difficulty for dualist interactionism is of course how something immaterial can affect something physical. But suppose that

that difficulty can be overcome, perhaps along Humean lines. Then a further problem arises which has not been much noticed. Descartes (*Passions* §34) said that the *slitghtest* movement in the pineal gland may alter the course of the animal spirits *very greatly*. So if the soul were to make small errors in the movements it transmits to the pineal gland, these might get amplified disastrously; how does avoid it such errors? We may call this the "sure touch" problem. It arises equally for Eccles. He wrote: "Presumably the self-conscious mind does not act on the cortical modules with some bash operation, but rather with a slight deviation. A very gentle deviation up or down is all that is required" (Popper and Eccles 1977, p. 368). It is good that the mind is not a brain-basher but treats the brain gently. But how is the mind able to get its "very gentle" deviations just right?

To make it specific, consider the speech of a small child whose soul has not been in liaison with its brain for very long. When the child asks *"Where those dogs goed?"* the vocal noises sound all right to its ears (and to ours; it's the grammar that we don't find quite right). Small children don't seem to commit the vocal equivalent of a typo, inadvertently emitting *"Where fose bogs koed?"* or whatever. How are they able to have this nice control over larynx, lips, tongue, etc.? I now have quite a nice control over my word-processor, but in the early days I made all sorts of mistakes and spent a lot of time consulting guide-books and pestering friends for advice. But there are no guide-books or counsellors to tell the child's soul just which cortical modules to play upon, and just what gentle deviations to give them. What explanation could the divine origin hypothesis offer for our child's ability to emit right-sounding vocal noises?

The suggestion that a newly arrived soul sets about acquiring this sort of know-how by trial and error runs up against the enormous complexity of the cerebral cortex. Imagine yourself a newcomer to a space ship's control cabin, where you are confronted with 10,000 controls each with 10,000 positions; you can be quite sure you will never acquire control of the ship through a trial and error process. But by Eccles's estimates the cerebral cortex consists of something of the order of 10,000 modules each consisting of something of the order of 10,000 neurones (pp. 228, 242). The remaining possibility allowed by the divine-origin hypothesis is that when a soul is sent into a body it is already divinely pre-programmed. But this means that before entering the body the soul, instead of being the simple, pure substance traditionally assumed by adherents of the divine-origin hypothesis, is dependent on antecedent natural processes, though at one remove. If He is to pre-program a soul appropriately before sending it into a body, God will need a detailed plan of the naturally evolved brain to which the soul is to be pre-attuned; for instance, to prepare the soul for control over the mechanics of speech He would presumably need, among other things, an exact map of Broca's area. The pre-attunement hypothesis turns God into a middleman between biological evolution and human mentality.

Armed with the Spearhead Model Mk II, one can say: *"Je n'ai pas besoin de cette hypothese"*: there is no need for a Third-Party to prepare the mind for control over the body if they have been evolving together, and moreover the control system has been leading the way with the motor system following in its wake. True, the dualism of control capacity and motor power is not the same as the dualism of mind and body, and the Spearhead Model on its own says nothing about the extent to which a hominid's conscious processes are part of its control system. But the evolutionist theory within which the Spearhead Model was developed provides a cogent argument for "the efficacy of the mental." It runs as follows. Major premise: *if* a phenotypic character appears on the evolutionary scene relatively late and subsequently spreads widely among a great variety of species, becoming more strongly pronounced during phylogenetic development, *then* there is an overwhelming presumption that it is being selected for its survival value. Minor premise: consciousness satisfies the antecedent clause in the above premise. Lemma: something can have survival value for an animal only if it affects its bodily performance. Conclusion: there is an overwhelming presumption that consciousness affects bodily performance.

True, a mental event's having efficacy does not guarantee it a place in the control system. Suppose that you feel a sudden stab of pain for which there is no obvious explanation. Then it goes away and you forget about it. A month later this happens again, and you again forget about it. But when it happens a third time you decide to see your doctor. In this case we might say that the pain had a certain efficacy – it helped to get you to the doctor's waiting-room. But it did so by nudging your control system from without, as it were. But now consider a champion billiard-player preparing to make a stroke. After surveying the table to select the most promising combination, he sights along his cue, mentally estimating the required collision angle and momentum, positions his body just right, and makes a nicely controlled shot-and the balls move just as he planned. He is exploiting skills developed by our hominid ancestors playing survival games rather billiards. In cases like this, thought processes surely are playing a crucial role within the control system whatever materialists and epiphenomenalists may say. Popper was surely right when he said that "the biological function of the mind is clearly closely related to the mechanisms of control" (1977, p. 114).

I will conclude by bringing the Spearhead Model, reinforced with the assumption that hominids' minds are part of their control systems, into conjunction with the well known fact that there has been an astonishing threefold increase in the size of hominid brains over the last three million years. Animal's brains being, quite obviously, central to their control systems, this implies a tremendous enlargement of hominids' control capacity. There have also, of course, been significant if less dramatic developments on the motor side. Let us try to imagine for a moment that it had been the other way round: a runaway increase in motor power accompanied by relatively modest developments on the control side. The Spearhead Model tells us that could not happen, since the rapidly expanding motor system would have been getting more and more out of control. It would be rather as if one had put on seven-league boots which enormously increased one's stride, only to discover that one can control where one's foot lands only within a wide margin. But a runaway development on the control side is entirely consistent with the Spearhead Model, at least in its Mk II version, which as we saw allows evolutionary development to rise way above the *OA* line. And the Spearhead Model suggests that with a development of this kind the motor system is coming under an ever better, more precise, and effective control. Speech mechanisms exhibit this amenability to a highly precise control very convincingly. I mentioned earlier a small child effortlessly hitting the right notes, as it were, in saying *"Where those dogs goed*?" and avoiding *"Where fose bogs koed*?". And a normal adult will never inadvertently say "abreast" instead of "addressed" despite these words' close phonetic resemblance. Human fingers sometimes exhibit a quite astonishing amenability to control; think of a concert pianist sailing seemingly effortless through a Chopin mazurka or whatever. The case of Houdini is particularly relevant here. He said: "I have to work with great delicacy and lightning speed," adding that he had had "to make my fingers super-fingers in dexterity, and to train my toes to do the work of fingers" (quoted in Kellock 1928, p. 3). That he was able to get his *toes* working for him in skilled ways is a rather striking corroboration of the implication of the Spearhead Model in conjunction with the fact of hominid brain enlargement, that we have large reserves of control capacity. All of which suggests that evolution has endowed us large-brained creatures with more possibilities of autonomy, or self-rule, than we had supposed.

London School of Economics
Emeritus Professor until his death in 1999.

REFERENCES

Bartley, W.W. III (1987). Philosophy of Biology versus Philosophy of Physics. In: Radnitzky and Bartley III (eds.), *Evolutionary Epistemology, Theory of Rationality and the Sociology of Knowledge*, pp. 7-46. La Salle, IL: Open Court.

Dawkins, R. (1982). *The Extended Phenotype*. Oxford: Oxford University Press.

Descartes, R. (1646). *Passion*. In: *The Philosophical Writings of Descartes*. 3 vols. Translated by John Cottingham, Robert Stoothoff and Dougal Murdoch. Cambridge: Cambridge University Press, 1988.

Fisher, R.A. (1930). *The Genetical Theory of Natural Selection*. Oxford: Clarendon Press.

Hardy, A. (1957). Discussion: Following H. Graham Cannon's Paper: What Lamarck Really Said. *Proceedings Linnean Society* **168**, 85-87.

Hardy, A. (1965). *The Living Stream*. London: Collins.

Kellock, H. (1928). *Houdini: His Life Story*. London: Heinemann.

Popper, K.R. (1934). *Logik der Forschung*. Vienna: Springer. English translation in Popper (1959).

Popper, K.R. ([1945] 1966). *The Open Society and its Enemies*. 2 vols. London: Routledge and Kegan Paul. 5th ed.: 1966.

Popper, K.R. (1957). *The Poverty of Historicism*. London: Routledge and Kegan Paul. Originally published in *Economica* **11** (1944); **12** (1945).

Popper, K.R. (1959). *The Logic of Scientific Discovery*. London: Hutchinson. English translation of Popper (1934) with new preface, footnotes and appendices. 3rd ed.: 1972.

Popper, K.R. ([1963] 1972). *Conjectures and Refutations*. London: Routledge and Kegan Paul.

Popper, K.R. (1972). *Objective Knowledge: An Evolutionary Approach*. Oxford: Clarendon Press.

Popper, K.R. (1974). Autobiography of Karl Popper. In: Schilpp (ed.), pp. 3-181.

Popper, K.R. (1975). The Rationality of Scientific Revolutions. In: R. Harré (ed.), *Problems of Scientific Revolution*, pp. 72-101. Oxford: Clarendon Press.

Popper, K.R. (1978). Natural Selection and the Emergence of Mind. *Dialectica* **32**, 339-355.

Popper, K.R. (1982). The Place of Mind in Nature. In: R.Q. Elvee (ed.), *Mind in Nature*, pp. 31-59. San Francisco: Harper and Row..

Popper, K.R. and J.C. Eccles (1977). *The Self and Its Brain*. New York: Springer International.

Radnitzky, G. and W.W. Bartley III, eds. (1987). *Evolutionary Epistemology, Theory of Rationality, and the Sociology of Knowledge*. La Salle, IL: Open Court.

Schilpp, P.A., ed. (1974). *The Philosophy of Karl Popper (The Library of Living Philosophers)*. 2 vols. La Salle, IL: Open Court.

Mario Bunge

SEVEN DESIDERATA FOR RATIONALITY

Many of us take pride in being rational animals, but only philosophers have attempted to find out what rationality is (or what 'rationality' means). However, for better or for worse they have not reached a consensus. It is for the better, because lack of consensus may indicate that the problem is still open. And for the worse, because dissensus may indicate confusion, which is often a result of superficiality.

I submit that a root of our problem is that the word 'rationality' stands for at least seven different concepts. I shall also argue that these concepts and the desiderata they entail, are ordered in a certain manner, and that singling out any of these desiderata, at the expense of the others, leads to partial rationality, an easy prey to irrationalism. Therefore I shall make a plea for global or scientific rationality. Finally, I will suggest that Popper and Chomsky, who pass for being the contemporary champions of rationalism, profess only a bounded rationalism. I shall also suggest that paraconsistent logic and decision theory, for all their mathematical apparatus, are pseudorational. All this will be done sketchily: the subject is large and complex enough to deserve a systematic and exact book-length study.

1. Seven Concepts of Rationality

I shall distinguish the following concepts of rationality:

(i)   *conceptual*: minimizing fuzziness (vagueness or imprecision);

(ii)   *logical*: striving for consistency (avoiding contradiction);

(iii) *methodological*: questioning (doubting and criticizing) and justifying (demanding proof or evidence, favourable or unfavourable);

In: E. Suárez-Iñiguez (ed.), *The Power of Argumentation* (*Poznań Studies in the Philosophy of the Sciences and the Humanities*, vol. 93), pp. 117-128. Amsterdam/New York, NY: Rodopi, 2007.

(iv) *epistemological*: caring for empirical support and avoiding conjectures incompatible with the bulk of the body of scientific and technological knowledge;

(v) *ontological*: adopting a consistent world view compatible with the bulk of the science and technology of the day;

(vi) *valuational*: striving for goals which, in addition to being attainable, are worth being attained;

(vii) *practical*: adopting means likely to help attain the goals in view.

Mathematicians and logicians excel at conceptual, logical, and methodological rationality. Scientists are supposed to abide by methodological, epistemological, and ontological rationality. Technologists, managers, and men of action are likely to stress valuational and practical rationality. But most of us do not stick consistently to any of the seven rationalities, and few if any value all seven. Those who do will be called "full rationalists."

Philosophical rationalists have stressed the first three types of rationality. Value theorists, moral philosophers, and the occasional social scientist (in particular Max Weber) have focused on the last two (*Wertrationalität* and *Zweckrationalität*). On the other hand epistemological and ontological rationality, though often practiced, have been typically overlooked by rationalist philosophers. Moreover, what I have called "epistemological rationality" usually goes by the name of "the principle of empiricism." I suggest that this is a misnomer, for it is one thing to look for empirical support, and another to postulate that experience is the only source of knowledge – which happens to be the first axiom of empiricism. One may care for experience while being a (global) rationalist, just as one may study behaviour without being a behaviourist. The difference between rationalists *lato sensu* and empiricists, with regard to experience, is that between "tap" and "top": whereas the former demand that experience be always on tap, and reason always on top, empiricists wish to reverse these roles.

2. The Rationality System

The various concepts of rationality distinguished in the last Section are not mutually independent but form a system. In fact, they are ordered in the way they occur in the list included in that Section: they constitute a partially ordered set. We proceed to prove this claim.

Firstly, logical rationality, or internal consistency (or non-contradiction), presupposes conceptual rationality. In fact, the laws of any logic proper hold only for exact concepts and propositions, and fail for imprecise ones. (This rules out fuzzy "logics" as logics proper.) If C is vague, so is not-C, ergo the extensions of C and not-C have a non-empty overlap, whence C does not satisfy the principle of non-contradiction. (Consequently, different students of one and the same inexact thinker are bound to propose not just different, but also mutually incompatible, "interpretations" of the master's imprecise statements, none of which is necessarily the more faithful to the original.)

Secondly. methodological rationality presupposes logical rationality. In fact, doubting presupposes logical rationality: a proposition p can be subject to doubt only if it is logically possible for not-p to be inconsistent with p. Likewise, proving or disproving involves some formal logic or other, explicit or tacit, ordinary or deviant, as long as it includes the principle of non-contradiction or some equivalent of it. Likewise justifying and disproving involve some formal logic or other, explicit or tacit, ordinary or deviant (as long as it includes the principle of non-contradiction). In fact, giving genuine reasons for or against a hypothesis h amounts to producing a set of propositions that entail h, or that follow from h or that conjoined with h entail previously admitted propositions. Just as consistency is the hub of logical rationality, entailment is that of methodological rationality.

Thirdly, epistemological rationality presupposes methodological rationality. To begin with, a datum e can be relevant to a hypothesis h only if, in addition to sharing some referents, e and h are either mutually compatible or incompatible – i.e., if it is methodologically rational to uphold both of them or to reject at least one of them. As for the demand that every conjecture be externally consistent, i.e., compatible with the bulk of the scientific and technological knowledge of the day, it is nothing but a demand for overall consistency – i.e., logical rationality – as well as for full epistemological rationality. In short, epistemological rationality calls for systemicity (best attained by building hypothetico-deductive systems) as well as for empirical support. (The fact that overall consistency is hard if not impossible to attain, for we often employ mutually inconsistent theories, is beside the point. The point is that epistemological rationality is a desideratum. More in Section 4.)

Fourthly, ontological rationality presupposes epistemological rationality. In fact, a unified world view must be consistent with our best knowledge of the natural and social world: if it were not it would violate epistemological rationality. Now, of all the logically consistent world views only one is clearly compatible with factual science and technology, namely naturalism (or materialism). In fact, neither science nor technology countenance any of the entities typically postulated by the various forms of idealism, such as sensations or ideas in themselves (distinct from brain processes), much less supernatural agencies. Moreover, the admission of either self-existing ideas or of deities would set limits upon inquiry, technological design, and rational action: it would curtail the domain of rationality by declaring certain entities and processes to be beyond the reach of rational knowledge or control. In short, ontological rationality calls for naturalism and disqualifies idealism, supernaturalism, and all of the eclectic mixtures of naturalism with idealism or with supernaturalism – such as Cartesian dualism and Popperian trialism.

Fifthly, valuational rationality presupposes both epistemological and ontological rationality – if it did not we would reach for impossible goals, such as getting something for nothing. Or we would strive for goals that, though attainable, should be shunned for requiring oppression or murder.

Sixthly, practical rationality presupposes valuational rationality. If it did not we would try to devise impossible means to attain impossible goals. In particular, doing good actions presupposes knowing what is good for the recipient. Remember the shipments of powdered milk to Africans and Asians who cannot digest milk for lack of a certain enzyme.

In conclusion, the seven types of rationality distinguished in Section 1 are related by the relation of presupposition. Since this is a partial order relation ($\leq$), our set of rationalities constitutes a partially ordered set. And this is about the simplest kind of conceptual system one can think of.

3. Desirability

Even admitting the need for distinguishing seven different desiderata for rationality (Sect. 1), and granting that they form a system (Sect. 2), it is possible to regard any or even all of them as undesirable, unattainable, or both. Let us presently look into the desirability problem; the attainability problem will be tackled in the next Section.

We begin by noting that neither of the seven desiderata for rationality, with the possible exception of the seventh, is intrinsically or absolutely desirable: each is desirable for some purpose or other. (Such relativity holds for nearly all values.) Thus, rationalists value conceptual precision, in particular exactness, because they wish to understand and to secure logical rationality. The pursuit of exactness for its own sake is often a barren academic exercise. Witness much of exact philosophy.

Nor is logical rationality an end in itself: madmen and speculative philosophers have been known to contrive consistent systems and to carefully justify their mad postulates. We value logical rationality for two main reasons. One is that contradiction generates an unlimited number of arbitrary propositions, relevant or irrelevant, true or false: *ex contradictoriis quodlibet.* (Note the similarity with cancer.) We also value logical rationality as a means for attaining methodological rationality, i.e., as a help in identifying and investigating problems. (If new information contradicts an accepted hypothesis, and we care for logical rationality, we shall study the problem, and either reform the hypothesis or revise the new datum.)

In turn methodological rationality, in particular testability is a means for epistemological rationality. In fact, asking for proof or for disproof, for positive evidence or for counterexample, is to call for the assistance of propositions other than the one under examination. Epistemological rationality stimulates the building of systems and of bridges between them, and discourages epistemic isolation – the mark of crackpot theories.

Epistemological rationality has an instrumental value: it makes for ontological rationality, which is to be sought unless one is engaged in purely conceptual investigations, for in such a case we may disregard the real world. The skepticism involved in methodological and epistemological rationality is necessary discipline, not nourishment. We live, and often die, by belief, not by criticism. Our ultimate cognitive goal is to build systems of justified (though still fallible) propositions, i.e., propositions that can be believed (*pro tempore*). In short, epistemological rationality is a means for attaining ontological rationality. (Correspondingly, epistemology, whether descriptive or normative, should be in the service of ontology.)

So much for the cognitive rationalities. But all of us, whether scholars or journeymen, have noncognitive goals as well as cognitive ones. Now, a value system can be rational or not, according to whether it is, or fails to be, consistent and whether it rests, or fails to rest, on some body of relevant and sufficiently true knowledge. Hence, from an axiological viewpoint ontological rationality is valuable for helping us attain valuational rationality. Put negatively: a crazy ontology can only inspire a crazy value system.

In turn, valuational rationality is to be sought because it helps us attain practical rationality. And practical rationality is valuable for helping us reach whatever practical goals we may have. This is where the buck stops in ordinary life.

To sum up, all seven rationalities are desirable. Moreover, we should procure them in the order indicated in Section l.

4. Attainability

Irrationalists deny that the rationalities we have discussed can be attained or, if attainable, are worthwhile. Half-hearted rationalists care only for a few rationalities, and skeptics only for conceptual and logical rationality. I submit that all seven rationalities are attainable as well as desirable, and I take this to be the thesis of *full* or *global rationalism.* To prove that all seven rationalities are attainable we need the only principle of modal logic worth knowing, namely Aristotle's axiom: "If *p,* then *p* is possible."

Conceptual and logical rationality are the daily fare of logicians and mathematicians. No one can deny this, but some scientists and philosophers believe that exactness and consistency are unattainable outside formal science. The evidence for this pessimistic thesis is dual: our mathematical models of real things are at best approximately true, and few if any factual theories are known to be internally consistent. This is true, but the rationalist can rejoin that every conceptualization is susceptible to improvement, and that we need not worry about inconsistency as long as we have not found a contradiction.

Of course, this optimistic thesis cannot be proved. Conceivably there are (unknown) unsurpassable limits to exactification. And it is also conceivable that nobody may detect some contradictions lurking in the most serviceable theories. However, *if* there are limits to exactification, we have not reached them. And *if* good theories are found to contain contradictions, there are enough clever people to patch them up while we wait for some geniuses to replace them with far better theories. The optimistic thesis concerning conceptual and logical rationality is taken for granted. Usually in a tacit manner, by any researcher bent on improving the theoretical knowledge he has inherited. In other words, the conceptual and logical rationalities are presuppositions of theoretical research in all fields of science and technology.

Methodological rationality is an even more powerful motor of inquiry, for every investigation starts by questioning and ends up by finding proof, positive evidence, or disproof. (True, refutationists do not care for positive evidence. But everyone else does, and with good reason: weeding is ancillary to cultivation, not a substitute for it.) Methodological rationality is at work in science, technology, and some of the humanities, and it will remain in force in these fields as long as there are researchers, i.e., questioners and searchers for positive and negative evidence.

Epistemological rationality – the demand for empirical support and overall consistency – is taken for granted in all the branches of factual science and technology. However, the condition of overall consistency is not easily satisfied. In fact, occasionally we are forced to use, in one and the same piece of research, pairs of theories that are at odds with one another. For example, in order to calculate reaction rate constants, a theoretical chemist may employ both quantum and classical mechanics. However, he does so with a bad conscience and hoping that in future someone will be able to proceed consistently. (The latter method is called an *ab initio* calculation.) We all deplore such unclean procedures, we do our best to refrain from using them, and we hope that future advances will allow us to dispense with them. In short, epistemological rationality is often attainable. And when not attained, it remains an ideal and a driving force of inquiry.

Ontological rationality – the demand for a consistent world view – is even harder to attain, yet equally valuable. Surely we know of blatant cases of ontological inconsistency, such as that of the physicist who holds that whatever happens, or fails to happen, at the microphysical level is the result of some measurement act; or the neuroscientist who claims that the mind is immaterial and acts on the brain. Yet once in a while we meet ontologically rational individuals, such as Galileo and Spinoza, Darwin and Einstein. Moreover, it is always possible to cleanse scientific theories in such a manner that no ontological inconsistency remains in them. For example, special relativity and quantum mechanics can be formulated axiomatically without any reference to observers, and psychological processes can be explained, at least in principle, as brain processes.

It is hardly disputable that valuational rationality is hard to attain, if only because valuation is often emotionally charged or rooted in vested interests. However, valuational rationality is occasionally attained. For instance, it is common experience that we often strive for realistic and worthwhile goals, such as staying alive and helping others do the same. But it is also true that we frequently reach for mutually incompatible goals, and that value theorists have not been very successful in helping us pick sets of values or even in checking the consistency of value systems.

Nevertheless, the point is that valuational rationality is occasionally attained.

Finally, practical rationality is less problematic. It is the driving force of the crafts and technologies. To be sure, it is not free from problems: sometimes a rational technical solution creates unforeseen difficulties. This is bound to happen to non-systemic i.e., piece-meal, technical proposals, such as that of improving the health care system without at the same time raising the standard of living and the level of education. Practical rationality can only be attained through planning of a very special kind, namely systemic (which takes the whole society into account) and democratic (which includes public participation). No doubt this is an ideal, but one that seems feasible. In any case, our point was far more modest, to wit, that practical rationality is occasionally attained.

5. Partial Rationality: Popper and Chomsky

Karl Popper has been hailed as the foremost rationalist philosopher of our time, and Noam Chomsky as the founder of rationalist (as opposed to empiricist) linguistic theory. Let us examine both claims.

No doubt, Popper comes close to being a rationalist. Nevertheless, there are several strong non-rationalist streaks in his philosophy. Thus, his conceptual rationality is limited by his refusal to elucidate his key concepts and by his contention that questions of meaning are worthless. His logical rationality is limited by his claim that the acceptance or rejection of protocol statements is largely a matter of convention. His methodological rationality is limited by his refusal to investigate problems of the "What is X?" kind and by his disparaging of positive evidence. His epistemological rationality is limited by his failure to admit that scientists evaluate their theories and procedures not only in the light of empirical evidence but also by the way they jibe with theories or procedures in neighbouring fields. Altogether, Popper's epistemology comes too close to skepticism for the rationalist's comfort, since it holds that we can be sure only of falsity.

Although Popper has not proposed a comprehensive ontology, he has sinned against ontological rationality on at least two counts. In fact, he has defended the interactionist version of mind-body dualism without bothering to elucidate the notion of mind-brain interaction. Nor has he acknowledged the very existence of a vigorous branch of psychology propelled by the monistic hypothesis that mind is identical to a collection of brain functions, namely physiological psychology. Nor has Popper attempted to square his methodological individualism with the fact that social science studies social systems rather than their individual components (who are the objects of biology and psychology). Moreover, his individualism is inconsistent for making use of the holistic and unanalyzed notion of social situation. On the other hand, there is no question but that Popper has favoured valuational and practical rationality, even though he has proposed neither a value theory nor an ethical system beyond Epicurus's maxim "Do not cause pain."

In conclusion, Popper is a half-way rationalist. Surely this is a great virtue in an epoch when most philosophers are only a quarter rationalists or even frankly irrationalists. But our point was just to prove that Popper's philosophy is not the paragon of rationality it is usually taken to be, whence the advancement of rationalism will involve going far beyond Popper.

Chomsky's work is another case of bounded rationality improperly advertised as rationalism *tout court*. While it has greatly increased conceptual and logical rationality in syntax, it has remained stagnant in other areas and has blocked progress in still others. For example, the notion of deep structure is still vague, and the school has yet to build a theory of meaning. The MIT school has sinned against methodological rationality in various ways, for instance by excusing exceptions as cases of defective performance (never of competence), and by baptizing language learning (as "LAD," or "language acquisition device") instead of promoting its empirical investigation. The claim that we are all born knowing universal grammar is another case in point, for nobody has ever produced the system of rules of such grammar, nor has been able to prove that it is indeed inherited.

Chomsky's doctrine violates epistemological rationality on several counts: it does not square with our knowledge of the (poorly organized) nervous system of the newborn; it ignores the social function and the social conditioning of language; and it relies heavily on intuition. Nor is the doctrine ontologically rational, for it claims that language is totally foreign to other systems of animal communication; that language is an exception to evolution; and that the production and understanding of speech are to be explained in terms of the mentalistic categories of prescientific psychology. (There is of course nothing wrong with holding, against behaviourism, that we can be in mental states. What is wrong is to pretend that one has explained a mental "faculty" by postulating a nondescript immaterial "mental structure" underlying the faculty: this is not explaining but redescribing.)

In conclusion, Chomsky's rationalism is severely limited. The rational aspect of his doctrine has greatly enriched linguistics but its irrational streaks have revived prescientific psychology and increased the isolation of linguistics from neuroscience and social science. It is safe to prophesy that Chomsky will be remembered for his scientific achievements and his moral integrity, not for his philosophical flaws.

6. Pseudorationality: Paraconsistent Logic and Decision Theory

Classical rationalism held tacitly that conceptual and logical rationality suffice. If this were true every consistent mathematical theory would be a triumph of rationality. But this is not the case: one can mathematize theology and build crazy mathematical models – of, e.g., immaterial ghosts or of a continuum of consumers in a perfectly competitive economy. Moreover, blatant irrationalism can lurk behind sophisticated symbolism. Paraconsistent logic is a case in point.

The peculiarity of paraconsistent logic is that the principle of non-contradiction is not a logically valid schema in it. Hence it would be the "logic" underlying inconsistent theories. Moreover, by (mis)interpreting logic in ontological terms, paraconsistent logic could be regarded as a formalization of dialectics, according to which all things are inherently contradictory. In particular, ultralogic, the most ambitious of all paraconsistent logics, would be about everything thinkable, conceptual or material, possible or impossible, rational or irrational. Thus, the classical rationalistic illusion that reason is self-sufficient ends up by destroying rationality.

Clearly, paraconsistent logics are non-rational by definition of logical rationality, namely because they do not include the principle of non-contradiction. And, being *a priori*, they cannot account for change, so there is no legitimate ontological rationality for them. Ordinary, not paraconsistent logic underlies all known scientific theories of change – physical or chemical, biological or social. One of the reasons for this is that, if we were to admit contradictions, the notions of truth and falsity would coalesce. Another reason is that no empirical datum could possibly serve as evidence for or against any theories. (Indeed if datum e were just as good as not-e, then if e supports theory T, not-e could not possibly undermine T.) A third reason for keeping the principle of non-contradiction is that it underlies practical rationality. Indeed, when interpreted in practical terms, contradiction leads to inaction, for it is impossible to do *A* while at the same time refraining from doing *A*. In short, for all its formal apparatus, paraconsistent logic is non-rational.

Another instance of pseudorationality is decision theory in its usual interpretation. This interpretation involves subjective values (utilities) and subjective probabilities (credences or belief strengths). Since subjectivity is epistemologically non-rational, the standard interpretation of decision theory is itself non-rational and, therefore, it is not an adequate tool for rational decision making. A fully rational theory of decision under uncertainty (hence under risk) should involve objective values and objective probabilities. Compliance with this epistemological requirement would ensure methodological rationality, for it would render the theory empirically testable. As it is, the defenders of the theory excuse its lack of empirical support claiming that it is normative, not descriptive. However, this line of defense is not acceptable, for a good normative theory is one that guides successful action – which is not the case with decision theory.

Nor is that the only reason that decision theory is pseudorational. Another is that it contains the principle that a rational decision maker must attempt to maximize his expected utility, by opting for either the "sure thing" (large probability, small payoff), or for the Quixotic adventure (small probability, large payoff). But this is not how responsible decision makers actually proceed. While they are willing to run some risks, they never risk ruin or death for the prospects of large payoffs – particularly when such prospects have been assessed intuitively (subjectively) rather than scientifically. Gambling is irrational. So is any theory that purports to be a guide to gambling.

In short, mathematization is often necessary to achieve rationality but except within pure mathematics, it is never sufficient. Worse, methodological, epistemological, ontological or valuational irrationality may lurk behind many a pretty mathematical formalism.

7. Concluding Remarks

Since there are different kinds of rationality, it is possible to satisfy or to violate some or all of them. We may call *semirational* any doctrine that satisfies some but not all rationalities, and *full rationality* any doctrine that satisfies them all. On the other hand *irrationalism* is the family of doctrines that reject all kinds of rationality. However, no irrationalist philosopher is fully non-rational if he argues cogently (though from false premises) against the desirability or attainability of rationality. (Kierkegaard, unlike Heidegger, is a case in point.)

Full rationality is a desideratum both theoretical and practical. Like many other desiderata, it is hard to attain. But it is attainable because its components are mutually compatible and, moreover, they constitute a system. Furthermore, full rationality is both theoretically and practically desirable because it guides successful inquiry and action.

This being the case, one may well ask why full rationality is so rare. A possible answer is that it is hard to attain. And it is hard to attain partly because it is extremely young, and partly because rational inquiry threatens every static doctrine, and rational action threatens every static institution. Consequently promoting full rationality involves not only investigating, arguing and teaching, but also fighting. Perhaps this is why Athena-Minerva wore a warrior's helmet.

McGill University
Department of Philosophy
855 Sherbrooke Street West
Montreal, Quebec H3A 2T7
Canada
e-mail: marioaugustobunge@hotmail.com

Ambrosio Velasco Gómez

THE HERMENEUTIC CONCEPTION OF SCIENTIFIC
TRADITIONS IN KARL R. POPPER

1. Introduction

Karl R. Popper is undoubtedly one of the authors who have contributed most to the development of the philosophy of science in the 20th century. His vigorous criticism of the central theses of logical positivism opened up new perspectives for the analysis of the rationality of sciences, that would subsequently be critically taken up and developed by authors such as Thomas S. Kuhn and Imre Lakatos.

One of Popper's ideas that had the greatest influence on Kuhn and Lakatos was the thesis that there is no firm, definitive empirical basis, but rather that, by convention, the latter must be accepted in a provisional, fallible way. Linked to this thesis is the idea that the rationality of science lies not in its empirical bases but rather in the way scientific theories are criticized, change and evolve.

If the rationality of science lies in its peculiar progressive historical change, then scientific methodology must act as a resource for promoting change rather than for seeking and proving the truth of scientific theories. It is therefore evident that method in sciences is not a method of verification, as the Positivists thought, but a method of empirical refutation.

Due to the fact that scientists seek to empirically refute their theories and hypotheses, it is possible to detect errors and overcome them through the formulation of bolder and better theories. This process of conjectures and refutations gives rise to the progressive historical transformation of sciences.

Popper describes historical change as the continuous development of the scientific tradition. Although the concept of tradition is crucial to

In: E. Suárez-Iñiguez (ed.), *The Power of Argumentation* (*Poznań Studies in the Philosophy of the Sciences and the Humanities*, vol. 93), pp. 129-141. Amsterdam/New York, NY: Rodopi, 2007.

Popper's philosophy, it has failed to receive the attention it deserves. This concept is so important that Popper himself was obliged to put forward a specific methodological proposal for the understanding of scientific traditions and social traditions in general. This methodological proposal actually constitutes one of Popper's most valuable contributions to social sciences, and one that also distinguishes him radically from the Positivists. Popper believed that his methodological proposal for social sciences, which he called "situational analysis," actually constituted a hermeneutic proposal. By vindicating a hermeneutic method, Popper moved radically away from the Positivists, who were extremely scornful of hermeneutic methods in the social sciences.

Popper's contribution to hermeneutics in the social sciences is another issue that has barely been explored in Popperian studies.[1] In this paper, I propose precisely to analyze Popper's contribution to the concept of tradition and hermeneutics. Highlighting these issues gives us a better understanding of Popper's significance, not only in the philosophy of science, but also in socio-historical sciences. In the following section, I shall focus on Popper's theory of the rationality of traditions. I shall then analyze his contributions to hermeneutics and subsequently emphasize the consequences of these contributions for an understanding of scientific rationality.

2. The Rationality of Scientific Traditions

Karl R. Popper put forward a set of theses that would subsequently be developed by post-Positivist philosophers of science with a certain claim to originality, such as Kuhn, Hesse, Shapere, Toulmin, Lakatos and Laudan. These theses include the conventional nature of the empirical base[2] and methodological rules, but above all the idea that the development of scientific knowledge is conditioned by historically inherited theories, methods, criteria and values, which orient the posing of problems, the formulation of hypotheses and their acceptance or rejection. Popper calls these inherited elements "tradition."

---

[1] One of the few works that analyze this issue is the article by Farr (1983).

[2] See Popper (1959, Chapter V). It is important to note that the conventionalist theory had previously been formulated by Otto Neurath in 1932. "It is impossible to take formal sentences conclusively as the starting point for sciences. There is no clean slate. We are like navigators who are forced to transform their ships in mid ocean, without ever being able to dismantle them in a sand dyke and reconstruct them with the finest materials available" (Neurath 1981, p. 206).

By recognizing the fact that tradition is a key element in the development of knowledge, Popper opposes the rationalist, empirical conceptions of modernity that regarded tradition as an obstacle to the objective basing of knowledge on indubitable sources (reason or experience).[3]

By opposing science and tradition, modern philosophers not only denied the historical fact that "we free ourselves entirely from the bonds of tradition" and that "the so-called freeing is really only a change from one tradition to another" (Popper 1963, p. 122).[4] In addition to ignoring this issue, in their desire to create a rational basis for knowledge, rationalist philosophers lost sight of two essential epistemological functions played by traditions, the first of which is to provide us with knowledge that has been created over centuries and milleniums. "If we start afresh, then, when we died, we shall be about as far as Adam and Eve were when they died." In order to be able to progress in science, Popper stated, "We must stand on the shoulders of our predecessors. We must carry on a certain tradition" (Popper 1963, p. 129). The second important epistemic tradition is the development of a critical attitude and skills that will enable the scientist and the philosopher to reflect on the givens of traditions in order to be able to free themselves from their prejudices and taboos. This liberation may be achieved, either through the reflexive acceptance of aspects of tradition or through their rejection and replacement with new contents.

By distinguishing two epistemological functions, Popper recognizes two types of intellectual traditions. On the one hand, there are the specific traditions comprising specific theories and myths, that transmit substantive knowledge developed by previous generations and which constitute "traditions of the first order." These traditions are the principal source of our knowledge. However, this does not mean that this type of tradition provides a rational justification for knowledge.[5] The task of

---

[3] Through this recognition of the epistemological importance of intellectual traditions, Karl R. Popper is partly accepting Michael Oakshott's criticisms of modern, contemporary rationalism. Indeed, Popper's formulation of the concept of tradition is a response to Oakshott's article, "Rationalism in Politics," originally published in *The Cambridge Journal* in 1947. In this study, in addition to questioning the universalist claims of modern, contemporary rationalism, Oakshott defends the importance of tradition in the sphere of moral and political knowledge (see Oakshott 1962).

[4] Karl R. Popper, "Towards a Rational Theory of Tradition" lecture given in 1949 and reproduced in his book (1963).

[5] "The fact that most of the sources of our knowledge are traditional condemns anti-traditionalism as futile. But, this fact must not be held to support a traditionalistic attitude: every bit of our traditional knowledge is open to critical examination and may be

justifying knowledge corresponds to another type of tradition, that is more of a transhistorical metatradition, lacking specific cognitive contents and consisting of a critical attitude and methodology. This "tradition of the second order" is the critical rationalism invented by Greek philosophers over two thousand years ago and one that has continued to be the most important feature of scientific knowledge, even today (see "Towards a Rational Theory of Tradition" in Popper 1963, p. 126). This critical tradition has made it possible to overcome traditionalistic, dogmatic attitudes and to live rationally in and through specific traditions that we are continuously transforming.[6]

Specific traditions of the first order and the universal (or transhistorical) tradition of the second order not only fulfill different functions (the source and the justification of knowledge) but also have very different characteristics. Traditions of the first order are plural, always emerge in specific historical contexts and change continuously as a result of the critical evaluation undertaken of them from the tradition of the second order. This last type of tradition is unique, invariable and of universal scope. Specific traditions of the first order are passive in that they merely provide material for critical evaluation. Conversely, the critical tradition of the second order is active, in that it undertakes a critical scrutiny of traditions, thereby fostering their change and progress. However, the critical tradition is not reflexive, does not subject itself to criticism and therefore does not change or evolve.

---

overthrown. Nevertheless, without tradition, knowledge would be impossible" ("On the sources of knowledge and of ignorance" in Popper 1963, section XVI, p. 28).

[6] The distinction between tradition and traditionalism that Popper establishes has been developed more clearly by Edward Shils in the social and political sphere: "*Traditionalism* is not only hostile to liberty, it is also radically hostile to *tradition*, the vague, flexible tradition that which even when it does not include the tradition of liberty does at least allows liberty to live in its margins of ambiguity to grow gradually and to take deeper roots. In oligarchical societies, traditionalism prevents the further growth of the elements which can give rise to freedom. In societies where liberty has already been established, traditionalism is the greatest enemy of civility . . ." (Shils 1997, p. 116). In the sphere of social theory and political philosophy, other authors such as Peter Winch, Michael Oakshott and Alasdair MacIntyre also reject the dogmatic nature commonly associated with traditions, while accepting the possibility that participants in a tradition may develop a critical, reflexive and rational attitude towards it.

These characteristics are summarized in the following chart:

Table 1.

Scientific traditions of the first order		Meta tradition of the second order
Substantive theories and inter-pretations	Contents	Critical attitude and methodology: Critical rationalism
Rooted in specific historical situ-ations	Scope	Transcends historical contexts and seeks universality.
Transmit knowledge that was produced and accepted in the past	Function	Critically analyze and evaluates inherited knowledge
Change continuously and pro-gressively as a result of critical methodology (may be falsified) and continuously progress.	Validity	Has not changed since Ancient Greece, does not question itself reflexively, is not falsifiable and does not progress.

The rational progress of science can be understood as a synthesis of the two types of traditions: Metatradition continuously questions specific scientific traditions as a result of which these traditions progressively evolve towards truth (verisimilitude), despite the fact that a scientific theory or tradition can never be completely true. Thus, truth in the sciences is gradually achieved through their historical development, while the critical metatradition does not in itself change or develop historically.[7]

3. Hermeneutics and Scientific Traditions

Popper did not restrict himself to proposing a philosophical model of scientific change in the terms I have just described. He also put forward a hermeneutic methodology for understanding changes in scientific traditions. Insofar as scientific traditions (of the first order) are historical, social events, Popper agreed that his study largely corresponded to socio-historical sciences. Consequently, Popper would have to accept that the philosophical understanding of the rationality of scientific traditions depends partly on the contributions of social sciences. It is perhaps for this reason that Popper was decidedly interested in the methodology of social sciences, particularly hermeneutic methodology. This interest led

---

[7] This static, ahistorical conception of critical rationalism as a metatradition was one of the main targets of criticism of authors such as Kuhn and more specifically Laudan. (See Laudan 1977, chapter III.)

Popper to radically distance himself from the positivist conceptions of the social sciences.

Contrary to the positivist conceptions of sciences, Popper accepted the fact that understanding was the goal of the social sciences and the humanities. Contrary to the hemeneutic conceptions of socio-historical sciences such as those of Dilthey and Collingwood, however, Popper believed that understanding was not exclusive to social sciences, or to the humanities, but was also the goal of all science.[8] In this respect, Popper criticizes both the positivists, who mistakenly sought to impose a natural science model on any disciplines, and the hermeneutic humanists, who uncritically accepted the fact that "positivism or scientism is the only philosophy appropriate for natural sciences" (Popper 1972, p. 174). In this respect, Popper appears to define a hermeneutic monism that was equally valid for natural and social sciences alike.

In addition to this uncritical acceptance of the positivist conception of natural sciences, Popper questions the subjective nature of empathetic understanding proposed by Dilthey and Collingwood. Instead of *Verstehen* or empathetic understanding, Popper proposed situational analysis as a method of objective understanding. This method involves reconstructing the "problematic situation" of the scientist or in general of any agent whose works require understanding. The problematic situation consists of the problem to which the theory seeks to respond and the "background" or "cognitive framework" in which the author raises the problem and attempts to resolve it.

> This background consists, of at least, of the *language* that which always incorporates many theories in the very structure of its usages and of many other theoretical asumptions, unchallenged at least for the time being. It is only against a background like this that a problem can arise. (Popper 1972, p. 165)

It is essential to stress the idea of background, precisely because it defines a set situation that enables one to raise certain problems, conjecture certain solutions and critically assess them. I believe that this notion of background may be replaced by the concept of tradition (of the first order) that Popper develops in other text (see "Towards a Rational Theory of Tradition" in Popper 1963, pp. 120-135). Background or specific traditions constitute an essential linguistic and theoretical framework for understanding the scientific theories of the past and for

---

[8] "I am prepared to accept the thesis that understanding is the aim of the humanities. But I doubt whether we should deny that it is the aim of natural sciences also" (Popper 1972, p. 185).

evaluating the author's rationality for proposing and defending them. Imposing theories or concepts from outside the author's context to evaluate the rationality of his theories and arguments always constitutes a "failure of historical understanding." Thus, the "method of situational analysis may be described as an application of the rationality principle" (see Popper 1972, p. 179) since it attempts to discover the reasons why a scientist accepted or rejected certain hypotheses within his particular context.

By proposing situational analysis as a method for understanding the rationality of actions and human products, Popper was obviously not implying recognition of the success of the action. Thus, for example, in the case of Galileo's theory of tides, which Popper recognizes as false from the outset, he shows the rationality underlying Galileo's decision to maintain the hypothesis of the Earth's circular movement. In contrast to Galileo's critics who accused him of being dogmatic, Popper pointed out that "Galileo's method was correct when he tried to proceed as far as possible with the help of the rational conservation law of rotary motions" (see Popper 1972, p. 174). Popper explained that the failure of Galileo's theory of tides was not due to a failure of rationality in his reasoning, but rather to a flaw in the framework or background of his particular problematic situation. Galileo was aware of the unsatisfactory nature of his theory, but inasmuch as he had no tradition of other theories and laws (such as forces of attraction, for example) in his background, he was unable to develop the solution that Newton subsequently formulated in another problematic situation.

I would like to point out here that background or tradition in the Popperian conception is something that transcends specific individuals and even determines or substantially limits the hypotheses and theories that they can formulate.[9] This inherited knowledge in one's background or tradition forms part of the third world or objective mind, in the Popperian sense. Insofar as it involves a world of autonomous cultural objects (theories, ideas, traditions, hypotheses, interpretations, etc.) rather than subjective events, depending on specific individuals, Popper cites "the superiority of the third world method – consisting of critically constructing problematic situations – over that of the second world, consisting of intuitively reviving a personal experience" (Popper 1972, p. 170, note 18).

---

[9] À propos of this point, it is interesting to analyze Popper's coincidences with H.D. Gadamer, since both acknowledge and value the hermeneutic situation and given prejudices in tradition. I have analyzed several points of coincidence in my article (Velasco 1997, pp. 125-145).

The superiority of the situational analysis method is based on the fact that the historical understanding of actions or works from the past does not require doubtful psychological skills on the part of the historian, but rather a reconstruction of historically identifiable cultural objects (objects from the third world), from which the problems, responses and arguments of the author or agent whose works are being studied are constructed.

Popper obviously does not ignore the important function of the author or agent in the maintenance and transformation of traditions, since although they define limits for posing problems and solving them, they are also an object of questioning, criticism and dissatisfaction on the part of the author, which promote changes and innovations (that are either imperceptible or revolutionary) within the same traditions that are handed down to them.[10]

Popper did not hesitate to state that is precisely the change in problematic situations, rather than the success of theories that determines the progress of the sciences.[11] In this respect, the most important function of theories is to raise new problems whose solution requires a change in the background or tradition from which they originally emerged. Hence, we can conclude that for Popper, the heuristic function of theories for discovering and formulating new problems is more important than the degree of corroboration of these theories. Consequently, although Popper does not explicitly acknowledge this, and it may appear paradoxical in relation to what he proposed in *The Logic of Scientific Discovery*, heuristics is a fundamental aspect in the context of justification. This statement anticipates an argument that began with Popper yet also went against him.

---

[10] " . . . The history of sciences must not be taken as a history of theories but rather as a history of problematic situations and their modifications (sometimes imperceptible, sometimes revolutionary) through the interventions of attempts to solve the problems" (Popper 1972, p. 177).

[11] "Thus we may say that the most lasting contribution to the growth of scientific knowledge that a theory can make are the new problems which it raises, so that we are led back to the view of science and of the growth of knowledge as always starting from, and always ending with, problems – problems of an ever increasing depth, and an ever increasing fertility in suggesting new problems" (Popper 1963, p. 222).

4. The Anti-Popperian Consequences of Popperian Hermeneutics

As we have seen, Popper does not regard understanding as being exclusive to social sciences and humanities but also to natural sciences. Inexplicably and unfortunately, Popper believes that the reconstruction of the problematic situation of the scientist is a meta-problem that the historian of science raises, but was not raised by the scientist whose work he seeks to understand.[12] This is even more disconcerting if we recall that Popper realized that the task of the historian was to reconstruct the objective problematic situation "just as it is." If the reconstruction of the problematic situation aims to be objective, it must necessarily resemble the situation faced by the author. Although one might accept, not without discussion, the fact that the agent had a false image or knowledge of his problem, it would be impossible to deny that the reconstructed situation sought to provide an accurate description of the scientist's problematic situation. Otherwise there would be no point in talking about understanding the rationality of the scientist's actions or works. It is therefore impossible to speak of two distinct, separate levels: the reconstructed and the original problems of the scientist.

If we take the task of situational analysis seriously, we must recognize that this method seeks to objectively understand the activity and products of scientists and therefore seeks to describe the problems, methodologies, arguments and theories put forward by the author in certain situations. In this respect, the interpretations that the hermeneutic method of situational analysis can provide cannot be different from the philosophical and historical interpretations that attempt to describe the scientific theories and investigations that are developed in specific traditions.

On the basis of this argument, we can therefore reformulate a Popperian characterization of scientific knowledge that is far removed from the one that Popper himself developed in *The Logic of Scientific Discovery* or in his polemic against Kuhn. This description is compatible with the hermeneutic theses of natural and social sciences developed, among others, by Kuhn and Lakatos in the Anglo-Saxon tradition and by Gadamer and Ricoeur in philosophical hermeneutics. I shall now put forward some of the theses I propose for this reformulation of the Popperian concept of science.

a) Any theory or interpretation must be considered a response to a set problem. Both the problem and the response are delimited by existing

---

[12] "We have to distinguish clearly between the metaproblems and metatheories of the history of sciences and the problems and theories of scientist" (Popper 1972, p. 177).

conceptual, methodological and theoretical resources, as well as by the debates and commitments in force in the background or tradition to which the author of the theory belongs.

b) The rationality of a theory must be evaluated on the basis of existing resources in this background or tradition. In particular, it is necessary to focus on the limits defined by the problematic situation that enable one to provide certain answers and exclude others. In this respect, it is not possible to have recourse to supposedly universal criteria, methodologies or standards, but only to those available in the specific circumstances of the tradition in question.

c) The validity of a theory or interpretation must not only be defined in terms of the success of a theory in relation to the existing resources and criteria in the tradition or background, but also in relation to the failures and limitations that this tradition imposes, limitations that cannot only be described by the historian but by the scientist also.

d) The tension that arises between maintaining what is given in the tradition or background and change and innovation goes beyond the methodologies and criteria of validity in force in this tradition. The decision over what parts of the tradition or background should be left and what parts rejected cannot be a demonstrative, methodical decision, either in the sense of falsification (rejection of the given) or corroboration (recourse to and recognition of the given). In any case, it involves a prudent decision, based on practical rationality.[13]

e) Questioning the criteria of universal rationality and validity leads to recognition of their contextual and historical nature. However, this recognition does not lead to radical relativism, but merely to affirming that, just as theories change and progress, so do our criteria and methodologies.

f) The progress of any theory, interpretation, tradition or methodology is primarily linked to its ability to pose new problems that will reveal (or construct) new aspects of reality (whether natural or social). The heuristic potential, (the capacity for discovery), is the principal criterion for evaluating our theories, interpretations, methodologies and standards of rationality.

---

[13] This idea was clearly formulated by Pierre Duhem in 1906: "Since logic does not determine with strict precision the time when an inadequate hypothesis should give way to a more fruitful assumption, and since recognizing this moment belongs to *good sense*, physicists may hasten this judgment and increase the rapidity of scientific progress by trying consciously to make *good sense* within themselves, more lucid, and more vigilant" (Duhem 1977, p. 217).

5. Final Comments

The two concepts on which our analysis of Popper's philosophy of sciences have focussed have been *tradition* and *hermeneutics*. They are certainly concepts that are not normally emphasized in the interpretations of the work of Karl R. Popper, perhaps because they belong more to Continental than to Anglo-Saxon philosophy. Yet it is precisely this circumstance that demonstrates Popper's philosophical audacity in transcending the limits of his tradition and learning from foreign traditions in order to question and transform his own philosophical tradition.

The interpretation I have just provided of Popper's philosophy of science indicates and indeed radicalizes the critical and innovative aspects of Karl R. Popper, to such an extent that they present an "improper Popper." In other words, a Popper who appears not to see himself reflected in the traditional image of the author of *The Logic of Scientific Discovery*, and who appears to be much closer to his critical rivals such as Kuhn and Laudan.

Yet this crisis of philosophical identity is an advantage rather than a drawback in Popper's thought, since it shows that his work not only recovers and develops the major achievements of logical positivism and the methodological conception of scientific rationality, but also inaugurates and postulates new conceptions of the rational development of science and obviously new perspectives for its understanding.

The most distinctive feature of this new conception of scientific rationality consists precisely of conceiving it as a process of the historical transformation of scientific traditions. This process consists precisely of the critical and creative use of the theoretical resources that scientists inherited from their predecessors for raising new problems whose solutions require the transformation of inherited knowledge. Thus, innovation, discovery and the heuristic force of hypotheses and theories constitute the determining factor in the rational progress of scientific traditions.

As we have seen, this historical understanding of the rationality of traditions requires a hermeneutic method, i.e., situational analysis. Popper thereby acknowledges the fact that the philosophy of science is primarily a historical discipline, a hermeneutic discipline. This anticipates the conclusion that Thomas S. Kuhn would reach many years later regarding his own work on the history of science.

The most important aspect of Popper's hermeneutic work, however, is his conception of the nature of scientific rationality: it is a sensible, practical rationality that goes beyond the narrow confines of methodology. If we accept this prudent conception of scientific rationality, many of the objections raised to the Popperian conception of science will therefore prove unfounded. It would therefore, for example, dispel Feyerabend's criticisms of the centrality of method, since scientific method would not be a fetish to which the scientist was subjected but rather a resource that could be used sensibly. It would also serve to counter Lakatos and Kuhn's criticism of the rationality of the permanence and tenacity of theories that would apparently be untenable on the basis of a strictly falsifiable methodological criterion. As we have seen, the perspective of prudent change recognizes both the rationality of change and the permanence of theories.

However, the advantage of the interpretation of Popper I have given here is not restricted to presenting a "stronger" Popper to counter historians' and philosophers' criticisms. Nor is it limited to depicting a Popper who anticipated many of the central ideas of the post-Positivist philosophy of science. Moreover, the conception of prudential rationality we have highlighted offers an alternative proposal involving a common rationality for the sciences, ethics, and politics.[14] This common rationality not only opens up new routes for a broader, more coherent interpretation of Popper's philosophy of science and political philosophy, but also offers new lines of analysis for the link between science, ethics and politics. And it is precisely these issues that lie at the heart of the debates on the philosophy of science at the beginning of the third millenium. For this reason, I believe that the hermeneutic reformulation of the Popperian conception of science that I have offered opens up new opportunities for transforming and developing the philosophy of Karl R. Popper towards new horizons, rather than merely interpreting his past achievements.

Universidad Nacional Autónoma de México
Facultad de Filosofía y Letras, Dean
04510 México
Distrito Federal
México
e-mail: ambrosio@servidor.unam.mx

---

[14] For a broader interpretation of the various philosophical spheres of the work of Karl R. Popper, see Suárez-Íñiguez (1998).

REFERENCES

Ayer, A.J. (1981). *El Positivismo Lógico.* Mexico: F.C.E.

Duhem, P. (1977). *The Aim and Structure of Physical Theory.* New York: Atheneum.

Farr, J. (1983). Popper's Hermeneutics. *Philosophy of Social Science* **13** (2), 157-177.

Laudan, L. (1977). *Progress and Its Problems.* Los Angeles: University of California Press.

Neurath, O. (1981). Proposiciones Protocolares. In: A.J. Ayer (ed.), *El Positivismo Lógico*, pp. 205-214. Mexico: F.C.E.

Oakshott, M. (1967). Rationalism in Politics. In: *Rationalism in Politics and Other Essays*, pp. 1-58. London: Methuen.

Popper, K.R. (1959). *The Logic of Scientific Discovery.* London: Hutchinson.

Popper, K.R. (1963). *Conjectures and Refutations. The Growth of Scientific Knowledge.* London: Routledge and Kegan Paul.

Popper, K.R. (1972). On the Theory of Objective Mind. In: *Objective Knowledge: An Evolutionary Approach*, pp. 153-190. Oxford: Oxford University Press.

Shils, E. (1997). *The Virtue of Civility.* Indiannapolis: Liberty Fund.

Suárez-Íñiguez, E. (1998). *La fuerza de la razón.* Mexico: Nueva Imagen.

Velasco, A. (1997). Universalismo y relativismo en los sentidos filosóficos de tradición. *En Diánoia: Anuario de Filosofía* **43**, 125-145.

David Miller

PROPENSITIES AND INDETERMINISM

1. Prefatory Remarks

In these prefatory remarks, which are designed to locate my topic within the complex and wide-stretching field of Popper's thought and writings, I shall not say anything that those familiar with his work will not already know. Moreover, what I do say will take as understood many of the problems and theories, not to mention the terminology, that I shall later be doing my best to make understandable. My apologies are therefore due equally to those who know something about Popper's discussions of indeterminism and of the propensity interpretation of probability, and to those who know nothing.

The Postscript to *The Logic of Scientific Discovery* was eventually published in three volumes: *Realism and the Aim of Science* (1983), *The Open Universe* (1982a), *Quantum Theory and the Schism in Physics* (1982b). It had been conceived in the first half of the 1950s as a series of new addenda to the parent work, but had grown at such a pace that the decision was eventually taken to separate it from its source and prepare it for independent existence. *The Logic of Scientific Discovery* (1959), the English translation of the original *Logik der Forschung* (1934), made its own way, appearing in 1959 still burdened with new addenda (twelve of them in the English edition, increasing to no fewer than twenty in the 10th German edition of 1994). But The Postscript did not follow, and was eventually withdrawn at galley-proof stage, to be subjected to a strenuous process of rewriting. As so often happens, new urgencies emerged, and by the early 1960s the project was as good as shelved. Only in about 1980 did the late Bill Bartley – who had been considerably responsible for the postponement of the book's parturition – undertake to prepare it for publication. And so, like its precursor, it first appeared in English

In: E. Suárez-Iñiguez (ed), *The Power of Argumentation* (*Poznań Studies in the Philosophy of the Sciences and the Humanities*, vol. 93), pp. 143-169. Amsterdam/New York, NY: Rodopi, 2007.

some 25 years after its origination, and like its precursor heavily encumbered with additional material.

When *The Logic of Scientific Discovery* was published in 1959, the anonymous reviewer [now identified as Kathleen Nott] in *The Times Literary Supplement* described it as a "remarkable book", and declared: "One cannot help feeling that if it had been translated as soon as it was originally published philosophy in this country might have been saved some detours." A strikingly similar remark is made by Magee in the preamble to one of his published conversations with Popper (Magee 1971, p. 66). I doubt that the same is true of *The Postscript*, at least to the same degree, since a fair number of its central themes were to appear in Popper's writings throughout the 1960s and 1970s,[1] often developed substantially beyond their *Postscript* versions, and others seem to have been introduced into public discussion by colleagues, several of whom (including myself for one weekend) had had access to the proofs. Nonetheless, it is unfortunate that the book appeared as late as it did. By the 1980s, some parts of it, especially parts of volume I, (Popper 1983), seemed dated, fighting battles long since won (either convincingly or by default) against opponents long since retired. The propensity interpretation of probability, the thread running through all three volumes, had been appropriated and articulated by others, and – if I may say so – perhaps misunderstood by some. Most seriously, some of the book's most challenging ideas found themselves overshadowed byindependent developments. I am thinking here especially of the attack in volume II (Popper 1982) on scientific determinism, which anticipated (but came after) the current fascination with non-linear dynamics. In this last case the situation looks at first glance to be a somewhat peculiar one, since the standard view is that the theory of chaos lends credibility to determinism, whereas Popper used very similar ideas to cast doubt on determinism. Although the situation is not as paradoxical as that makes it sound, it does deserve to be looked at, and to be resolved.

The two topics of my title "Propensities and Indeterminism" are indeed closely interconnected in the original version of *The Postscript*. The propensity interpretation of probability, it is made clear there, makes limited sense – more exactly, it has limited interest – in a world strictly controlled by deterministic laws. Moreover, one of the central arguments used against metaphysical determinism (the argument known as Landé's blade) also illustrates the need for an objectivist interpretation of probability deeper than the frequency interpretation, and the propensity

---

[1] Compare the assessment of *The Postscript* on p. 455 of O'Hear (1985).

interpretation certainly has some title to be able to fill that role. (No matter that many, myself included, doubt whether Landé's blade is a sound argument. For if it fails as a criticism of determinism, it fails also as an illustration of the potency and usefulness of the propensity interpretation.) But it is not only in *The Postscript* as originally conceived that there is a close and intimate liaison between propensities and indeterminacies. In later papers, especially the 1965 lecture "Of Clouds and Clocks", reprinted in *Objective Knowledge* (Popper 1972), and the aptly entitled addendum "Indeterminism Is Not Enough" (Popper 1973), which helps to pad out (1982a) (Volume II of *The Postscript*), Popper maintained that for human freedom and creativity to flourish it is not enough that we live in an indeterministic physical universe. In addition, the physical universe must be causally open to influences from outside, not only mental activity but also abstract influences such as arguments and critical discussions. Even later, especially in the lecture *A World of Propensities* (1990), he came to stress the idea that, just as the propensity interpretation is meagre fare without indeterminism, so indeterminism is meagre fare without full-blooded propensities. In section 5 I hope to be able to provide some flavour of the enriched diet of propensities + indeterminism.

No single lecture could deal adequately with the two topics of my title, and I shall make no more than incidental references to the work of other philosophers. Even the entirety of Popper's own writings against determinism is beyond my power to survey. But it would be wilfully negligent to disregard altogether the central arguments of *The Poverty of Historicism* (Popper 1957), to the effect that there are no laws of development (let alone progress) in history, and that no scientific prediction of the course of human history is conceivable. Let me mention briefly the most celebrated of these arguments: the argument that the growth of scientific knowledge is unpredictable (Popper 1982a, section 21), and that since the course of human history depends decisively on the way in which scientific knowledge grows, human history is unpredictable too (the Preface to Popper 1957).

For all its popularity, this two-stage argument, it seems to me, is not quite satisfactory (which is not to say that its conclusion is not correct).

In the first place, as Popper admits, its intermediate conclusion may be false; it may indeed be the case (as some contemporary physicists appear to think) that "the growth of our theoretical knowledge has come to an end" and that all that is left is "the endless task of applying our theories to ever new and ever different initial conditions" (Popper 1982a, p. 67). Moreover, our inability to advance further theoretically may itself be one of those things that our present theories allow us to predict. That is, we may be able to predict from what we know now that the future growth of theoretical knowledge will be nil. More seriously, however, the second stage of the argument is invalid. For the unpredictability of a cause does not always mean the unpredictability of its effect. (All deaths are more or less predictable, though their causes often are not.) Science fiction demonstrates that uncannily accurate (if not very precise) predictions of the advance of applied science, especially technology, need not depend on any prediction of the appearance of any new scientific theory. (Science fiction is indeed mostly technology fiction, and genuine science fiction would be just a kind of – perhaps too easily falsifiable – science.) Against this it may be claimed that no scientific, rather than clairvoyant, prediction of an effect is possible without knowledge of its cause; to which the only correct retort is that technological advance is not in this sense an effect of scientific advance. It is natural to suppose that, if this is so, it is because much new technology is no more than an application of already existing science, but this response misses the point that what scientific theories tell us is what cannot be achieved, not what can be (Popper 1957, sections 20, 26), so that new theories alone cannot be the source of new technology.[2] In other words, even if we were to predict correctly the future of theoretical science, we should hardly have begun to be able to predict its impact. Much more important for any rational prediction of the future of technology is knowledge of what needs to be accomplished. Having said this, I at once acknowledge that new theories may bring to our attention previously unknown phenomena – such as the untapped energy resources within the atom – that can be exploited if we are inventive enough. It would be foolish to claim that the advance of theoretical science has no significance for the advance of technology. I want to stress only that it is not the primary source, and that the applications of science may be more predictable than science is itself.

To be sure, such predictions (like nearly all predictions in social life) will be rather imprecise, predictions in principle rather than predictions in detail. But the dogma of historicism that Popper attacked in (1957) was likewise concerned with the broad sweep of human history, rather than with its intimate details. Determinism is different. In its most provocative form it holds that the future of the world is predictable, at least in principle, in uncompromising detail. To this doctrine we now turn.

---

[2] The point is elaborated in Miller (1994), Chapter 2.2g and in Miller (2006), Chapter 5.3.

2. Determinism and Deterministic Theories

Determinism and indeterminism, which are mutual contradictories, may be amongst those philosophical theories – another is realism – that it is much easier to have an opinion on than to have a decent formulation of. Popper begins by listing a variety of deterministic positions: religious, scientific, metaphysical determinism. The last of these is characterized as the view that (Popper 1982a, p. 8)

> . . . all events in this world are fixed, or unalterable, or predetermined. It does not assert that they are known to anybody, or predictable by scientific means. But it asserts that the future is as little changeable as is the past. Everybody knows what we mean when we say that the past cannot be changed. It is in precisely the same sense that the future cannot be changed, according to metaphysical determinism.

In many respects this is a clear statement. In fact, if it were to be presented together with an idealistic theory of time, according to which the future is just as real or unreal as is the past, it would be unexceptionable. The trouble is that it is simply unbelievable as it stands. It appeals to the commonsense view that the past is out of bounds, yet it is yoked to a theory that is starkly counter to common sense. Common sense undoubtedly regards time as a real process, and maintains that the future, even if determined, does not possess the same kind of established (if inaccessible) reality that the past possesses. There is a difference between past and future that is incontrovertibly relevant to the claim that the past cannot be changed; and that is that the past has happened and the future has not. Russell suggested that "but for the accident that memory works backward and not forward, we should regard the future as equally determined by the fact that it will happen" (Russell [1918] 1953, p. 190), but the kind of logical determination at issue here is, as he conceded (Russell [1918] 1953, pp. 191f), less than the whole story. The image of the history of the universe as already developed on a cinematographic film, in short, the idea of a block-universe, is indeed a common concomitant of metaphysical determinism. Still, films pass through projectors, and this passage is a real process even if everything to be seen on the screen is fixed before the process begins. I should have thought that it should be possible to be a determinist yet to hold that the future is not yet realized (see also Popper 1982a, sections 11, 18, and 26).

Be that as it may, the doctrine of metaphysical determinism – which Popper says "may be described as containing only what is common to the various deterministic theories" (1982a, p. 8) – is plainly a metaphysical doctrine in the sense that it is not open to empirical falsification. No observational or experimental results, however little anticipated, can demonstrate that what is observed to take place was not fixed prior to the observation or the experiment. Nor can they demonstrate the opposite. Neither metaphysical determinism nor metaphysical indeterminism (the doctrine that some events, perhaps not many, are not fixed), that is to say, is open to direct scientific investigation. Now it is certainly no part of Popper's position in (1982a) (as it perhaps was in Popper 1934) to restrict rational discussion or investigation to the theories of empirical science – and perhaps also those of logic and mathematics. The determinism/indeterminism debate may be seen as a salient test case for the contrapositivist view that metaphysical theories, if formulated clearly, can be rationally discussed and criticized; that the modes of rational discussion are not limited to empirical research and logical and mathematical analysis. It is this concern that leads Popper to the formulation of what he calls "scientific determinism" (the topic of the next section). Let us see how what emerges.

One way in which metaphysical theories may be discussed is in terms of the problems that they solve, or were designed to solve but fail to solve (Popper 1963, Chapter 8). This works well for Greek atomism, to take a shining example, but it seems much less effective for determinism and indeterminism. For these doctrines seem more like attempts to evade problems than genuine attempts at solutions. Determinism does little more than generalize from the undoubted truth that many aspects of our lives evince vivid regularities and unexpected repetitions, regularities that it is not in our power to alter; the residual problem, simply deferred or legislated away, is how to account for the irregularities that permeate the regularity, for the endless novelty whose irregular but not infrequent appearance we almost take for granted, and for those regularities that we do seem to be able to alter and even to insist on. With indeterminism it is partly the other way round: the clouds are comprehensible, but not the clocks. It is part of Popper's thesis in (1982a) that there is indeed an asymmetry here; or rather, that there are two asymmetries. In the first place, indeterminism does not suggest that no events are fixed in advance, only that there are some that are not (1982a, p. 28). As Earman points out, this irenic doctrine may not be as innocuous as it seems (Earman 1986, pp. 13*f*), which does not imply that indeterminists are not free to espouse some kind of partial determinism (Popper 1982a, pp. 126*f*), and to explain clocks as more or less deterministic systems more or less isolated from external influence. Secondly, Popper claims (1982a, section 28), indeterminism (coupled with the propensity interpretation of probability) has some ability to explain statistical regularity, whereas determinism is impotent to explain irregularity (rather than just postulate it as an accumulation of unrelated effects). Were this latter claim true, there might be a case for supposing that indeterminism had a problem-solving edge. But nothing that I know at present gives me much optimism that it is true.

There is another way, almost a scientific way, in which metaphysical theories may be evaluated, and that is by revealing them as logical consequences of scientific theories, hence true if those theories are true. If suitable scientific theories are not available, they may have to be invented; it was by this process that the bald existential hypothesis "There exist neutrinos" eventually became incorporated into science. Is there then any way in which either metaphysical determinism or metaphysical indeterminism could be distilled from any scientific theory?

Once we pose the question in this form, we see that the matter is not straightforward. These doctrines, as formulated above, are concerned with the fixedness or unfixedness of the future, with what is or is not alterable. Scientific theories are not formulated in this kind of vocabulary.[3] This helps to explain how it is that determinism and indeterminism are both irrefutable. It cannot be true that each is, as Popper says of determinism, "irrefutable just because of its weakness" (1982a, p. 8), since the contradictory of a weak theory is correspondingly strong; indeed, it is the relative weakness of indeterminism that he usually stresses elsewhere (1982a, p. 28). It might be better to say that determinism, being universal, is a strong theory, but nonetheless an irrefutable one because not properly concerned with matters to which empirical investigation has direct access.

No scientific theory, as normally understood, logically implies metaphysical determinism in the formulation above, for no theory says that what it says will happen has to happen. Now perhaps the fault here is with the way that science usually formulates its theories, paying little attention to its own metaphysical heritage and ruthlessly eliminating metaphysical and modal elements. But even those who emphasize the metaphysical content of science distinguish between substantial metaphysics and idle metaphysics, metaphysics that does not contribute to testability or empirical content (Watkins 1984, p. 205). Yet it does not seem that by rewriting a scientific theory so that its predictions come garnished with necessity we make it magically more testable. The only

---

[3] Compare Russell's remark that "in advanced sciences such as gravitational astronomy, the word 'cause' never occurs" (Russell [1918] 1953, p. 171).

way of showing that something is not bound to occur is to show that it does not occur.

Classical celestial mechanics is presented by Popper as a prime example of what he calls a *prima facie* deterministic theory; that is to say, it is a theory that (1982a, p. 31)

> . . . allows us to deduce, from a *mathematically exact* description of the initial state of a closed system which is described in the terms of the theory, the description, *with any stipulated finite degree of precision*, of the state of the system at any given future instant of time.

(Complete precision cannot be expected, because some families of equations are soluble only by approximation methods.) But even if classical mechanics satisfies this condition, which is a more controversial claim than Popper seems to have appreciated (Earman 1986, Chapter 3), it does not imply metaphysical determinism. If N is classical mechanics, I an appropriate package of auxiliary hypotheses and initial conditions relevant to some isolated physical system, and f a description of some future state of the system, then of course $N \wedge I$ logically or necessarily implies f. But for familiar reasons it may be concluded that f is unalterable or necessary only if it is assumed that both N and I are unalterable. We may be quite prepared to make these assumptions, especially with respect to initial conditions I that relate entirely to events in the unalterable past, but it must be recognised that they are additional assumptions and not parts of classical mechanics.

It may be felt that these considerations give too much weight to the emphasis on unalterability that appears in the intuitive statement of determinism, and that it is enough to call the world deterministic if its behaviour is entirely described by a *prima facie* deterministic theory together with statements of initial conditions. There are well known perils in this approach, which we escape only if we are careful not to count any deductively closed accumulation of sentences as a theory (Russell [1918] 1953, pp. 192-194; Earman 1986, Chapter 2.5). If we suppose that we can do this – it seems to require some distinction between lawlike statements and others, and therefore some concession to the demand that the future of a deterministic world is bound by law, unalterable – we may be tempted to put meat on Popper's suggestion that metaphysical determinism contains "only what is common to the various deterministic theories" (1982a, p. 8). But it is by no means obvious that the outcome (the intersection $\vee \mathfrak{D}$ of the class $\mathfrak{D}$ of all *prima facie* deterministic theories) would be a recognisable formulation of determinism. Although it is trivial that *prima facie* deterministic theories have something in common, it is not obvious that they share any non-trivial logical consequences. The failure is plain in the complementary case of indeterminism. Since many *prima facie* indeterministic theories, if not all, have *prima facie* deterministic extensions, what is common to them all, the intersection $\vee \mathfrak{J}$, follows from many *prima facie* deterministic theories too, and can therefore scarcely be a formulation of indeterminism. (Indeed, it seems unavoidable that $\vee \mathfrak{J} = \mathbf{T}$, the class of logical truths.) What all this indicates is that there is no available statement of metaphysical determinism that does not make explicit reference to laws ($\vee \mathfrak{D}$ might provide such a statement, even though it is defined in terms of laws, but sadly we do not know what theory $\vee \mathfrak{D}$ is). The only formulation of metaphysical determinism that is implied by every *prima facie* deterministic theory seems to be the metalinguistic thesis that some *prima facie* deterministic theory is true. Metaphysical indeterminism must be stated as the negation of this thesis (rather than the thesis that some *prima facie* indeterministic theory is true).[4]

3. Scientific Determinism

By such means the doctrine of metaphysical determinism can be formally incorporated into science, provided that science contains a theory that is, as classical mechanics was supposed to be, *prima facie* deterministic. But it is plain that its position is peripheral and precarious, and that only if the theory is strictly true and quite comprehensive is anything said about whether determinism is true. This might not matter except that the comprehensiveness even of Newtonian physics has never been more than a dream (Popper 1982a, p. 38), and – given the pervasiveness of errors and approximations – the same might be said, and was said by Newton himself and by Peirce (Popper 1982a, pp. 212*f*), of its claim to be strictly true.[5] It would therefore be valuable to be able to formulate a deterministic position that digs somewhat deeper into science, and does not depend so crucially upon the existence of a virtually perfect theory.

---

[4] Compare Earman (1986, p. 13), who limits himself to distinguishing between deterministic and indeterministic worlds. As far as I can see, the only formulation of determinism given is that the actual world is a deterministic world. Much of Earman's book is devoted to the task of investigating whether the principal theories of classical and modern physics are *prima facie* deterministic.

[5] Cartwright (1983) and Forster & Sober (1994, pp. 1-35) are examples of contemporary writers who, not always for identical reasons, reject the view that real science is exact science.

A considerable part of (Popper 1982a) is concerned with a doctrine labelled 'scientific determinism'. Popper attributes the substance of this doctrine to Laplace (1819), who famously imagined (Popper 1982a, p. 30)

> . . . a superhuman intelligence, capable of ascertaining the complete set of initial conditions of the world system at any one instant of time. With the help of these initial conditions and the laws of nature, i.e., the equations of mechanics, the demon would be able, according to Laplace, to deduce all future states of the world system . . .

Scientific determinism is this idea of predictive potency cast in a testable form. Popper seems to intend it to be a scientific doctrine in at least three distinguishable respects:

(i)   like metaphysical determinism in its final formulation above (but not in its initial intuitive formulation), it concerns science, and makes explicit reference to scientific theories;
(ii) it restricts itself to real scientific capacities, in particular predictive capacities;
(iii) unlike metaphysical determinism, it is itself open to direct scientific investigation; that is, it is empirically falsifiable.

Let us take these points in turn.

(i) Like metaphysical determinism, scientific determinism is, for all its reference to scientific laws, a doctrine about the world, not about our knowledge of the world. Popper writes, for example, that "the fundamental idea . . . is that the structure of the world is such that every future event can in principle be rationally calculated in advance, if only we know the laws of nature, and the present or past state of the world" (1982a, p. 6). A little later he declares unequivocally that "[i]n asserting [scientific determinism] . . . , we assert of the world that it has a certain property" (1982a, p. 38).

(ii) Scientific determinism is stronger than metaphysical determinism, and says more than that the world is ruled by a *prima facie* deterministic theory (1982a, section 13), or that "every future event can in principle be rationally calculated in advance . . . " That would mean only that a completely precise description of the state of a system at an initial time could be transformed formally into a description (at any demanded level of precision) of its state at a later time. Scientific determinism says in addition that the formal mathematical relation in the theory that subsists between the earlier description and the later one can in principle be realized in a physical process (not merely in a mathematical calculation): the process begins with the collection of information appropriate to the formulation of the description of the system at the earlier time, and terminates in the formulation and publication of a prediction prior to the later time. Since this process is a physical one, the initial information cannot be expected in a perfectly precise form, though no finite bound to its precision need be insisted on. In short, the predictability that scientific determinism adds to metaphysical determinism is supposed to be what Popper calls "predictability from within" (1982a, section 11). It must be feasible, not just possible, and must make use of real physical processes.

(iii) Scientific determinism will not itself be a scientific doctrine if it says only that, given suffciently precise initial conditions, a prediction may be obtained at any preassigned level of precision; for failure to achieve a satisfactory prediction could always be blamed on an insuffcently precise starting point. It is therefore necessary to require that the *prima facie* deterministic theory with which we have equipped ourselves should satisfy also what Popper calls the principle of accountability. In its most appropriate form this says that, given any prediction task (1982a, pp. 12*f*, emphasis suppressed),

> we can calculate from our prediction task . . . the requisite degree of precision of . . . the results of possible measurements from which the initial conditions can be calculated . . .

That is, for any level of precision demanded for the prediction, we may work out in advance, with assistance from the theory, how precise we need to make our measurements of those quantities from which we calculate the initial conditions.

Popper offers two versions of scientific determinism satisfying the requirements (i)-(iii) (1982a, pp. 36*f*):

(I)   the state of any closed physical system at any given future instant of time can be predicted, even from within the system, with any specified degree of precision, by deducing the prediction from theories, in conjunction with initial conditions whose required degree of precision can always be calculated (in accordance with the principle of accountability) if the prediction task is given.

(II) whether or not a closed physical system will at any given future instant of time be in any given state can be predicted, even from within the system, by deduction from theories, in conjunction with initial conditions whose required degree of precision can always be calculated (in accordance with the principle of accountability) if the prediction task is given.

The stronger version (II) is of historical significance since it subsumes the problem, by which Laplace was exercised, of whether classical mechanics ensures that the solar system is dynamically stable. It is the version of scientific determinism that is refuted by considerations from non-linear dynamics.

A number of things could be said about the details of these formulations of scientific determinism. Here it suffices to note that not only the precision of predictions, but also their accuracy [closeness to the truth], needs to be mentioned if scientific determinism is to be a statement about the world, as intended in (i), rather than a statement about our theories about the world. I shall take this amendment for granted. About Laplace's demon, which he takes to be a forerunner of scientific determinism, Popper writes (1982a, p. 30):

> The crucial point about this argument of Laplace's is this. It makes the *doctrine of determinism a truth of science rather than of religion.* Laplace's demon is not an omniscient God, merely a super-scientist.

and a few pages later (p. 33):

> The crucial point is that . . . [scientific determinism] appeals to *the success of human science,* such as Newtonian theory: 'scientific' determinism is to appear as a result of the success of empirical science, or at least as supported by it. It appears to be based upon *human experience.*

It is perhaps not wholly clear that in the formulations given above scientific determinism is a scientific theory in its own right, as intended by the manufacturer. For could not any failure to make accurate predictions of suffcient precision, once the required precision of the initial conditions has been assured, be attributed not to the failure of scientific determinism itself, but to the falsity of the attendant explanatory scientific theory (say, classical mechanics)? Although the answer to this question is of course affrmative, two points should be made. The first is that we ought to regard an explanatory theory as falsified only if we obtain a reproducible effect that contradicts it (Popper 1959, section 22). But a possibility is that by sharpening the initial conditions we obtain predictions of varying precision (and indeed, accuracy), in which event we would do well not to reject the explanatory theory, but to reject the deterministic suggestion that better outputs can always be obtained from better inputs. I shall return to this point in a moment. In the second place, determinism is not a doctrine merely about the possibility of calculating what initial measurements are required for the delivery of particular predictions; as noted in (ii), it says also that these measurements are physically obtainable and the predictions physically deliverable. It was for this reason that I followed Popper above in stating the principle of accountability in terms of measurements rather than initial conditions; indeed, he says that "'scientific determinism' requires accountability in the stronger sense . . . a theory which is . . . non-accountable in the strong sense would be one whose determinist character could in principle not be tested by us . . . it could not be used to support 'scientific determinism'" (1982a, p. 13).[6] Recurrent failure suffciently to refine measurement techniques or speed up prediction techniques would therefore count against scientific determinism unless, of course, the bare possibility of unknown techniques was appealed to. That would, it must be admitted, undermine (and perhaps even destroy) any claim to empirical status.

So understood scientific determinism may seem to be not only falsifiable but plainly false. It requires the possibility of measurements made at unbounded levels of precision, and – what is worse – computations and publications generated at unbounded speeds. Predictions of any complexity are, remember, required to be possible for any future instant. Scientific determinists can take little comfort in the thought that the closer the initial and final times are, the less laborious should the computation have to be; this may be so – it may not be – but in general the presentation of the results can be expected to consume more time than does their preparation. Doubtless scientific determinism could be modified to deal with this piece of trouble, without serious compromise either to its deterministic character or to its scientific status. After all, if publishable predictions for tomorrow are promised, we may cheerfully condone the practical inability of the predictor to tell us what to expect in the next picosecond. It is interesting that predicting devices are usually more at a loss in the long term than in the short term. But if the dissemination of the prediction is an ineliminable part of the task, then short notice can be bad news too.

The main arguments against scientific determinism, however, are of a more logical or analytical kind than these. Popper argues in particular against the possibility of self-prediction, on the grounds that the activity of predicting (which is a physical activity) unremittingly encroaches on and disturbs the state of the predictor, whose future is what the predictor is aiming to predict (1982a, sections 22*f*). I am unable to evaluate this complex argument on the present occasion, but let me note that the

---

[6] The distinction is not registered by Honderich, who concludes that "successful predictions within neuroscience are evidence for determinism, even overwhelming evidence". The version of determinism espoused by Honderich does not maintain that measurements of ever-increasing precision are possible – indeed, it seems to deny it – and accordingly it falls short of – or contradicts – scientific determinism (see Honderich 1988, Chapter 6.6, especially p. 356).

interference of the predictor with its own future is no mere reiteration of the familiar point that every act of observation disturbs what is being observed (see 1982a, note 1 on p. 35). Nothing in scientific determinism requires a predictor to be able to predict how a system would have evolved had the initial measurements not been made on it. What is asked is only that suffciently precise measurements can be performed to yield a prediction of the behaviour of the disturbed system. For this reason I am unconvinced that scientific determinism needs to be linked to a *prima facie* deterministic theory that is comprehensive, or even that it implies metaphysical determinism. Indeed, the very possibility of manipulating initial conditions at will suggests that scientific determinism presupposes metaphysical indeterminism.[7]

Let me turn to something widely agreed, by determinists and indeterminists alike: the fact that if classical mechanics (and most other non-linear theories) are true then scientific determinism is false. Although very briefly treated, this is one of the principal theses of *OU* (1982a). It is one of the principal consequences too of the theory of dynamical chaos. The diffculty for scientific determinism is that in many interesting theories an accurate prediction of any precision at all can be achieved only by requiring unlimited precision in (the measurements of) the initial conditions. In a section of less than two pages (1982a, section 14). Popper resurrects a result by Hadamard discussed penetratingly in 1906 by Duhem (1954, Part 2, Chapter 3.3), which makes just this point. Imagine projectiles being thrown on to an infinite surface. Then some may settle into closed orbits, while some disappear to infinity. Hadamard established that on some simple varieties of surface of negative curvature there is no way, short of absolute precision, in which the angle of launch can be arranged so as to ensure either that the projectile falls into a loop or that it flies off for ever. Between any two launching angles that lead to loops there are some that lead to infinity; and between any two launching angles that lead to infinity there are some that lead to loops. The stronger form (II) of scientific determinism stated above is thus false, even if classical physics is true. No matter how precisely the launch is managed (short of absolute precision), qualitatively different outcomes are possible.

A more familiar example is the logistic function

$$f(t + 1) = \lambda f(t)(1 - f(t)),$$

---

[7] See also section 2 pp. 103-144 of Miller (1996).

where $f(0)$ lies in the open interval $(0, 1)$ and λ is in $[0, 4]$. Here too scientific determinism in the form (II) is contradicted, except for low values of λ and special values of $f(0)$.[8] For example, if $\lambda = 4$ and $f(0) = 0.75$ then f remains constant for ever. But almost every other initial value of f leads eventually to an irregular, apparently random, curve. No amount of initial precision short of absolute precision will allow us to predict that f remains constant, though if $f(0) \approx 0.75$ we have only to discover that fact by a sufficiently delicate measurement in order to be able to predict that f will not remain constant. But (II) is infringed. On the other hand, the weaker version (I) of scientific determinism is not controverted, since for every i the value of $f(i)$ is a continuous function of $f(0)$.

Where do Hadamard's result, and the logistic function, and the impossibility of self-prediction leave scientific determinism? It is clear that a *prima facie* deterministic theory may be true, yet scientific determinism false. Not only do *prima facie* deterministic theories not imply scientific determinism, some of them imply the negation of scientific determinism. Nothing follows, as far as I can tell, about the possibility of a more local kind of determinism that equally "appeals to *the success of human science, . . .* [and] appears to be based upon *human experience*" (1982a, p. 33), and concerns only predictability (up to a certain level of precision) in prepared systems isolated from the predicting apparatus.

Where do Hadamard's result and the logistic function, where does the attack on scientific determinism, leave metaphysical determinism? It is clear that whether or not it implies metaphysical determinism, the falsity of scientific determinism in classical physics tells us nothing about the truth or falsity of metaphysical determinism. Earman regards scientific determinism as "such a wrong-headed conception of determinism" that its failure is without significance (Earman 1986, p. 9).[9] Others (for example, Hunt 1987) seem ready to conclude that metaphysical determinism has been vindicated. Popper himself maintains that, because indeterminism is what we would now call the default position, "the burden of proof rests upon the shoulders of the determinist" (1982a, p. 27). But it would have been much better to have said that the burden of

---

[8] For details consult almost any book on chaos, for example Ian Stewart (1989, pp. 155-164). A graph of the first 400 values of the logistic function for $\lambda = 4$, $f(0) = 0.75000000012$ may be found in Miller (1994, p. 155).
[9] It is on quite different grounds that Earman himself takes metaphysical determinism to fail in classical physics.

criticism, and of exposing one's position to criticism, rests on all parties to a dispute. By being prepared to replace the unfalsifiable doctrine of metaphysical determinism by a stronger scientific thesis, determinists certainly stuck their necks out (to employ a characteristic Popperian expression), and they lost the gamble. That does not show metaphysical determinism to be false, but it does serve to emphasize how exposed is its position within the body of scientific knowledge – far more exposed, I should have thought, than that of metaphysical theses such as the principle of conservation of energy (Earman 1986, p. 10), whose regulative role is undeniable (Meyerson 1930, Chapter 5). If our fundamental theories turn out not to be *prima facie* deterministic, then that is the end of metaphysical determinism. It is now almost a commonplace that this is what has happened with the advent of quantum mechanics. Science has shown metaphysical determinism to be false.

4. Landé's Blade

Rather than discuss quantum mechanics, which I have no qualifications to do, I want to look here at the central philosophical argument in Popper (1982a) against metaphysical determinism. It is an argument to the effect that there are phenomena that only indeterministic theories can explain satisfactorily.

One of the most persistent themes in Popper's writing on probability theory over sixty years has been the claim that statistical conclusions can be obtained only from statistical or (latterly) probabilistic premises, and not from deterministic premises. In Chapter 8 of (1959) – a book more deterministic than indeterministic – he singled out two central problems for any theory of physical probability. One, the problem of the falsifiability of probability statements, though important, need not detain us here.[10] The other, the fundamental problem of the theory of chance, as he called it (1959, section 49), is the problem of explaining the statistical stability witnessed in many otherwise disordered sequences encountered in science and in everyday life. How are we to explain the appearance, in the midst of disorder, of a very tightly controlled order? Sequences of tosses of a die on a well regulated piece of apparatus, of results on a roulette wheel, of incidents of accidents at a busy street intersection, of the appearance of various characteristics in successions of generations in genetic experiments, and so on, all show the most remarkable statistical

---

[10] It is the central topic of Part III of Gillies (1973). See also Miller (1994), Chapter 9.3.

stability. Not only is it possible to estimate relative frequencies of 6-*up*, or of *even*, or of head-on crashes, or of blue eyes, by considering the frequencies shown in finite sequences, but these relative frequencies characteristically settle down very rapidly near to their final values, and thereafter do not deviate much from those values. Provided the macroscopic conditions that generate the sequences are unchanged, the sequences of outcomes are found to be remarkably stable. "The tendency of statistical averages to remain stable if the conditions remain stable is one of the most remarkable characteristics of our universe" Popper writes (1990, p. 12). Here is a problem for determinism and indeterminism alike. For indeterminists – who may be tempted to regard such sequences as sequences of wholly undetermined events – it is the problem of explaining why any order at all should appear in the midst of disorder; truly disordered sequences might be expected to be just that, without sign of constancy. For determinists, the problem is rather one of understanding disorder at all; and, given that it does occur, of explaining its relatively constant features.

Traditionally indeterminism has been able to give an explanation of statistical stability by construing these undetermined events as events in the thrall of a probability distribution. In von Mises's version of the frequency interpretation of probability, to be sure, the explanation did not amount to a great deal. Since probabilities are defined only for sequences in which frequencies converge, statistical stability is not much more than another way of saying that the type of event in question has a probability of occurring. For this and other reasons Popper felt obliged in (1959) to abandon von Mises's axiom of convergence, and to show that it could be deduced from an appropriately invigorated version of the axiom of randomness or excluded gambling systems. Even so, the explanation is hardly a very deep one, since randomness is explained in terms of the persistence of frequencies under various forms of selection of subsequences. There are frequencies mentioned in the premises, and it is hardly to be wondered at that there are frequencies mentioned also in the conclusion. The question of why it is frequencies that converge, that it is frequency-statistics that are stable, is not handled in depth in the frequency interpretation. One of the virtues of the propensity interpretation of probability is that it offers a somewhat deeper explanation of statistical stability.

Probability as such is not my topic in this paper, and I do not intend to describe the propensity interpretation of probability in much detail. Two of its principal features are that, unlike the frequency interpretation, it ascribes probabilities sensibly to single events, rather than to types of events, and that it relativizes probability ascriptions to the whole current state of the universe, rather than to the immediate locality. The probability of the occurrence of 6-*up with this die thrown now* is fixed in general not only by the physical features of the die, and by the features of the apparatus being used to throw it, and its immediate surroundings, but by the whole state of the world. The probability is said to be a measure of the propensity of the world to develop into one in which the outcome of the throw of the die is 6-*up*. The propensity that is the probability is not a propensity of the die, or of the apparatus, but of the world. It is not, as in the frequency interpretation, relative to a reference class.[11] As so understood, probabilities can take values other than 0 or 1 only if the world is metaphysically indeterministic – that is, in this case, the probability of 6-*up with this die thrown now* lies strictly between 0 and 1 only if the outcome of the throw of the die is not fixed in advance; the world has a propensity, which is neither cast-iron necessity nor cast-iron impossibility, to develop in the way described. If propensities, so construed, are postulated to satisfy the axioms of the calculus of probability, then we may conclude, through the laws of large numbers, that in stable circumstances, in which the propensities of the events in question do not change, there is an overwhelming propensity for a longish sequence of throws to be statistically stable. Now again this explanation is not a magically deep one, since the probability axioms are plainly satisfied in the most trite manner by frequencies, so it is no great miracle that stable frequencies should emerge as the conclusion. Nonetheless it should be stressed that, despite many comments by Popper himself that suggest just this, the propensities we are concerned with in such activities as games of chance are not fundamentally propensities to yield frequencies but propensities to produce single events. There do indeed exist propensities to produce frequencies, but these are explained in terms of the more fundamental propensities.

The propensity interpretation of probability is inescapably metaphysical, not only because many propensities are postulated that are not open to empirical evaluation but, more importantly, because the idea of necessity, or of law, or of compulsion is built into the theory from the start in a way that, as we saw in section 1, it need not be built into *prima facie* deterministic theories. The modal element in the propensity interpretation cannot be dismissed as idle metaphysics, for without it there is no objective interpretation of single-case probability at all. Here,

---

[11] Hence the criticism of the propensity interpretation given in section 3.3 of Howson (1995) is not a criticism of the propensity interpretation.

as elsewhere, it is rash to suppose that non-trivial probabilities are as well behaved and uncomplicated as trivial probabilities are.[12]

Metaphysical indeterminism accordingly explains statistical stability by pointing to stable propensities. How does metaphysical determinism fare? The express purpose of the argument known as Landé's blade, which is discussed in sections 29f of (Popper 1982a), is to show that determinism can give no explanation of statistical stability other than a completely trivial one; statistical stability is simply postulated. There is, that is to say, no deterministic explanation of stability. This argument is really the only argument against metaphysical determinism itself to appear in (1982a), and Popper set some store by it. Later Watkins echoed this emphasis on the argument, and expounded his understanding of it at some length (see 1974, section 2.4). But some ten years later, in response to unsympathetic criticism from the late J.L. Mackie, he retracted it in the form originally given, declaring it invalid, and going so far as to suggest that since the argument "had not . . . attracted much notice outside the Popper circle, it might seem that a quiet burial, perhaps in the form of a brief note of retraction, is all that is called for." But Watkins did not content himself with quiet interment. In addition to stating why he thought the argument to be invalid, he proposed a variant of it that he considered to be valid, and indeed to be "perhaps the strongest argument, outside quantum mechanics, against physical determinism" (see Watkins 1985, pp. 13-19). Although I shall not have to time to consider the matter in detail, I should like to spend a few moments on this topic. I too think that the argument is invalid, and that Landé's blade no longer cuts much ice, but not quite for the reason that Watkins diagnoses. If I am right, then Watkins's variant of the argument is also invalid.

Landé's blade is an imaginary device consisting of a vertical chute down which billiard balls can be delivered on to the edge of a blade. Although both Popper (1982a, section 29) and Watkins, following Landé, pay some attention to the positioning of the blade so that the distribution of the deflected balls on the two sides is 50 : 50, this does not, in my opinion, have any bearing on the argument. All that matters is that the blade is not so skew that all the balls fall to the same side (or, indeed, miss it altogether). In these circumstances, experience tells us, we shall find that the distribution of balls to left and right, whatever distribution it is, is random and seemingly unpredictable, but statistically stable. How is this stability (not the precise distribution, which may depend on many

---

[12] On this point see Miller (1996), pp. 104f. For further discussion and defence of the propensity interpretation see Miller (1994), Chapter 9.6 and Miller (2007).

special factors) to be explained? Metaphysical determinism, according to Landé, has only one route to an explanation, and that is to postulate hidden asymmetries at each impact of a billiard ball with the blade. (This is indeed what we quite correctly do when asked to explain the statistical stability of throws with a die or with a coin.) Some of these asymmetries – small draughts, spots of grease, a bit of English – are left-inclining, some right-inclining, and they are themselves distributed in what we describe as a random manner. Each is causally sufficient (we may suppose) to determine which side of the blade the ball will fall. Now Landé simply accepts this answer, and proceeds to ask for an explanation for the statistical stability of the left-inclining and the right-inclining causes. Determinism seems to be in no better position than before, and to be compelled to postulate another earlier sequence of asymmetries that is also statistically stable, and so on. The argument ends only when it is conceded that appropriate asymmetries must have been introduced into the universe at the very beginning, magically attuned to yield statistical stability aeons later. Alternatively, determinism offers no explanation but a concomitance of accidental occurrences. Whichever line the determinist retreats to, the indeterminist seems to have much the better of the encounter. The explanation of stability in terms of propensities may not be profound, but it is not empty, and it is not mystical.

The trouble with this argument is that its conclusion is false. The theory of dynamical chaos provides us with hosts of examples of random (or, perhaps better, pseudorandom) sequences generated deterministically. It is important to appreciate that there is no pseudorandomness in the sequence of values of the logistic function f mentioned in the previous section; this sequence would very quickly fail a test for randomness, since (for example), if $f(t) = 0.10$ then $f(t + 1) = 0.36$. But if we consider only whether the value of $f(t)$ is greater or less than 0.5, we shall obtain a sequence that has every appearance of being random. Dynamical chaos appears to provide just the kind of deterministic explanation of statistical stability that the argument of Landé's blade declares to be impossible.

Why is the argument of Landé's blade invalid? To see this, let us take an excursion into science fiction and graft the logistic function directly on to the blade, supposing that $f(t)$ measures some time-dependent characteristic of the blade. Suppose too that when a ball strikes the blade at time t, the value of f changes (by some mechanism that need not concern us) to $4f(t)(1 - f(t))$. If at the time t of an impact $f(t) > 0.5$, then the ball is deflected to the right, while if $f(t) < 0.5$, the ball is deflected to the left. (For simplicity, suppose that $f(t) = 0.5$ is impossible.) It is apparent that this purely deterministic piece of apparatus will generate a pseudorandom sequence of outcomes.

There are, as the determinist rightly maintains, small asymmetries at the edge of the blade that are responsible for each deflection. But it does not follow that each asymmetry is the causal outcome of a significantly earlier asymmetry, and that determinism must postulate sequences of antecedent asymmetries in the distant past. The causal chains do not stretch vertically back in time parallel to the chute, as it were, but horizontally along the sequence of previous changes of value of f. The mistaken assumption in Landé's argument is the assumption that because the falls to one side or another of the blade are statistically independent, therefore they and their antecedent causes are also statistically independent. This is simply false. Determinism can explain a statistically independent sequence of outcomes as the rigidly deterministic consequence of a statistically dependent sequence of antecedent conditions.

In his discussion of Watkins's original treatment of Landé's blade, Mackie suggests that the error lies in confusing randomness (that is, statistical stability) in the initial conditions, and randomness (that is, indeterminism) in the laws of development, and that the determinist can easily live with the former, though not of course with the latter (see Mackie 1978, pp. 369f). Watkins seems largely to agree.[13] But that is to miss the gist of Landé's blade, which asks for an explanation of statistical stability in nature. Watkins goes on to assert that "[a] confluence of rigidly deterministic causal chains that are largely independent of one another may yield a chaotic collective result; but this does not mean that a determinist has to postulate a chaotic ancestral state of which the present chaos is the descendant" (Watkins 1985, p. 16); and in support of this assertion he gestures in the direction of gas dynamics to provide a counterexample. But he simply assumes at this point that gas dynamics (actually billiard ball dynamics, but it comes to the same thing) is deterministic. Since it was the purpose of Landé's argument to show that, because independent causal chains emanating from a starting point showing no statistical stability will not deterministically generate statistical stability, gas dynamics must be indeterministic, Watkins fails, this second time round, to come to grips with the central issue.

---

[13] He seems even to hold that the indeterminist too has to accept that "the set-up is beset by various little asymmetries and disturbing influences, but these do not have any systematic bias towards either left or right" (Watkins 1985, p. 16). This is surely a mistake. Indeterminism assumes at most the constancy of the propensity of balls to fall one way or another, and is unbothered by whether or not there are asymmetries present.

The reason that the argument of Landé's blade fails is that the determinist need not accept that independent outcomes are the progeny of independent acts of generation. But it hardly suffices for the determinist simply to note this and pass on. In cases where statistically stable outcomes are apparently produced from regular and non-random beginnings, we need details of what the non-linear process responsible is. My grafting of the logistic function on to the blade was, as I admitted, no more than science fiction. If the determinist cannot do better than that, then he has given us no more than a promissory note of determinism, and is maintaining an explanatory theory in a form truly beyond challenge. To be sure, in the case of a macroscopic chute and blade, as in the tossing of a die or a coin on to a soft surface, I accept, as everyone else does, that there are usually small asymmetries present in the apparatus (and, indeed, that these can with care be eliminated). But I do not think that this is obviously so in the case, say, of incidence of photons at a half-silvered mirror.

The upshot of this is that I remain sceptical of Watkins's attempt to provide a valid analogue of Landé's blade, in the case in which the sequence of outcomes is not a statistically stable string but an intentionally meaningful one such as a conversation. Why, Watkins asks the determinist, should the antecedent conditions "have been so nicely geared to . . . [the speaker's] needs one billion years later?" (Watkins 1985, p. 18). The answer that the determinist will offer to this is that there is no relevant sequence of antecedent conditions one billion years before the conversation. The relevant initial conditions are all there in the conversation itself. There is no sequence of independent labial and lingual movements in need of explanation by antecedent causes, but a plainly dependent sequence that could in principle be the outcome of some deterministic evolution from a single starting point. I have no disagreement with Watkins concerning the unsatisfactoriness of such a glib answer. Unfortunately I am unable to see how considerations similar to those of Landé's blade are able to denounce it.

5. A World of Propensities

Lying behind most discussions of determinism is an interest in human freedom and creativity. As his discussion in ([1966] 1972), sections vii-ix, makes clear, Popper is one philosopher who takes the deterministic threat very seriously, and is under no illusions that there would be any authentic freedom in a physical world that was fully determined. More, he has emphasized how important it is for us that the world should not only not be determined at the level of physics – in the domain that, from the early 1970s onwards, he referred to as world 1 – but that it should be causally open to other influences; especially those from world 2 – the world of mental activity – and (through the intercession of world 2) those from world 3 – the world of abstract human creations, especially problems and theories (Popper 1966, section x and 1972). Our theories, created by us and encoded by our efforts in structures in world 1, are thereafter sustained there without direct intervention from us. On being rethought, they are able to bring about effects in world 1 itself, for example the construction of new machinery to the designs of an abstract blueprint. But even this turns out not to be enough to make sense of human cultural achievement (in the broad sense of that term). We are not sophisticated machines responding to stimuli but, like all living things, problem-solvers attempting to make our way – and I really mean make our way – in the world. The existence of causal indeterminacies is essential to these attempts of ours, but a mere menu of abstract possibilities from which we may be served is not sufficient. At the very least the available possibilities must be endowed with some potential of realization, some activity. It is here, I think, that Popper saw some metaphysical rewards to be gained from the postulation of active propensities, which may be harnessed by us in order to drag ourselves forwards into the open future. For propensities give promise of providing the much needed "medium betwixt chance and an absolute necessity" (Hume [1739] 1888, p. 171, quoted in [1966] 1972, p. 227).

From the beginning of the enterprise of interpreting most physical probabilities as propensities of the world (or in special cases, parts of the world) to develop in certain ways Popper had likened these propensities to Newtonian forces. To be sure, they combine in very different ways from Newtonian forces, and they can be annulled in a way that Newtonian forces cannot be. The greater force does not always prevail.[14] But they are supposed to have the same kind of potentiality as these forces. In (1990), he tells us of the propensity interpretation that (p. 9)

> it was only in the last year that I realized its cosmological significance. I mean the fact that we live in a world of propensities, and that this fact makes our world both more interesting and more homely than the world as seen by earlier states of the sciences.

---

[14] This has been a standard criticism of the propensity interpretation of probability. See, for example, D.H. Mellor (1971, p. 158) and Anthony O'Hear (1980, pp. 136-137). The reader is referred to the text to note 5, above.

I think that the phrase 'in the last year' does less than justice to some of the speculations in "A Metaphysical Epilogue", Chapter 4 of (Popper 1982b), but let that pass. In any event, what we have here is an attempt to see the cosmos, including ourselves, as the result of the realization of propensities and of the emergence of new possibilities. This question of the emergence of novelty was dominant in Popper's thinking about indeterminism, to the extent that Bartley, the editor of *The Postscript*, was moved to promote the book under the general banner that something can emerge out of nothing, challenging the traditional wisdom that there is nothing new under the sun (see the Editor's Foreword to Popper (1982b), p. xiii.). We may take the following two successive paragraphs from pp. 18-19 of (1990) as exemplary:

> This view of propensities allows us to see in a new light the processes that constitute our world: the world process. The world is no longer a causal machine – it can now be seen as a world of propensities, as an unfolding process of realizing possibilities and of unfolding new possibilities.
>
> This is very clear in the physical world where new elements, new atomic nuclei, are produced under extreme physical conditions of temperature and pressure: elements that survive only if they are not too unstable. And with the new nuclei, with the new elements, new possibilities are created, possibilities that previously simply did not exist. In the end, we ourselves become possible.

I should like to end my discussion of Popper's views on propensities and indeterminism with a very brief look at the problem of how new possibilities can emerge, and in particular whether such emergence of novelty is compatible with the vision of the world as an all-encompassing field of probabilistic propensities. For at first sight it does not seem that there can ever be genuinely new possibilities.

There is one uncontroversial sense in which new possibilities can come into existence, even in a deterministic world. For what is possible at one time may not have been possible at an earlier time, and hence the passage of time itself may suffice to make possible what previously was not possible. On the day I was born it was impossible that I should give a lecture at the Universidad Nacional Autónoma de México. But this occurrence, once impossible, has now all but conclusively been shown to be possible. To take care of this, a determinist would insist that the occurrence claimed once to be impossible, later possible, must be provided with a date, though perhaps not a very precise one; and that once this is done it is clear that there always existed a possibility (at least since I was born) that I should give a lecture at the Universidad Nacional

Autónoma de México in November 1994. Those who talk about the emergence of new possibilities mean something more than that. Moreover, they mean something less than the emergence of new logical possibilities. No doubt there are some who would want to say that logic is only something human, and there is no reason why it should not expand or contract. But Popper was not among them, and I have no intention of pursuing the question in that direction.

What is surely meant in (1990) by saying that new possibilities can be created is not merely that new possibilities can be created, but that new propensities, new forces – or centres of force – can be created. The problem with this picture is that it seems to clash with the identification of propensities and probabilities. If a dated event has no propensity, and hence a zero probability of occurring, then that probability can be raised to a positive probability only by the occurrence of another event of no probability, or no propensity. This very familiar consequence of probability theory, that only events of zero probability can alter the probabilities of other events of zero probability,[15] is one that Popper himself wielded on a number of occasions against others (see, for example, 1959, appendix ∗ vii, especially p. 364). It is a little surprising to see his own ideas apparently at the mercy of the same result. For what it means is that new non-zero propensities cannot emerge without the earlier actualization of events of zero propensity; and this looks very much like saying that new possibilities can come about only if something impossible happens first.

Popper certainly intends zero probability to indicate the absence of propensity. He writes, for example, that "zero propensities are, simply, no propensities at all, just as the number zero means 'no number' . . . a propensity zero means no propensity" (1996, p. 13). But despite the close connection between propensity and possibility, there is no suggestion that zero probability implies impossibility. On the contrary, he writes elsewhere that "zero probability . . . means, in the case of random events, a probability which may be neglected as if it were an impossibility" (1983, p. 380). Herein, I think, lies the solution to our problem.

We must recognise not only that zero propensity does not imply impossibility, but that there exist all the time possibilities whose propensity to occur is strictly zero. Popper himself nudges us in this direction with the remark that propensities are *"more than mere*

---

[15] If $p(a, c) = 0$ then $p(ab, c) = 0$ by B1 and (18) of Popper (1959), appendix ∗ v. By M2, $p(a, bc)p(b, c) = 0$. Thus b can raise above 0 the probability of a, given c, only if b's own probability, given c, is 0.

possibilities" (p. 12). Such events have no propensity to occur, but they may occur nonetheless – by accident, as it were. Indeed, accidental or chance occurrences seem to me to exactly what we need here. To misuse an example of Aristotle's, a tile slides off a roof and strikes a passer-by. Two unrelated causal chains, as we put it, coalesce by chance. There was no propensity at all for such an occurrence, but it was not abstractly impossible. It is the kind of occurrence that immediately brings into being new propensities, propensities that were previously zero. Serendipity is more significant than is sometimes thought. Like Popper, I find incredible the idea that at the beginning of the universe, or after the first three minutes of it, there was any propensity for *Don Quixote* eventually to be written, or for *The Naked Maja* to be painted, or for Teotihuacán to be built, even though these achievements were abstractly possible. If we admit that some things may happen without there having been any propensity for them to happen, we can happily accept that for most of recorded and unrecorded history there were simply no such propensities.

The world is therefore not a world run wholly by the operation of propensities. It contains many chance events too, events for which there never was any propensity of occurrence. It is these chance events that somehow we have learnt to take advantage of. I need hardly insist that there has to be much more to the story than this.

University of Warwick
Department of Philosophy
Coventry CV4 7AL
United Kingdom
e-mail: d.w.miller@warwick.ac.uk

REFERENCES

Earman, J. (1986). *A Primer on Determinism.* The University of Western Ontario Series in Philosophy of Science, vol. 32. Dordrecht: D. Reidel Publishing Company.

Cartwright, N. (1983). *How the Laws of Physics Lie.* Oxford: Clarendon Press.

Duhem, P.M.M. (1954). *The Aim and Structure of Physical Theory.* Princeton: Princeton University Press. English translation of *La Théorie Physique, son object, sa structure.* 2nd edition. Paris: Riviére & Cie, 1914.

Forster, M. and E. Sober (1994). How to Tell when Simpler, More Unified, or Less *Ad Hoc* Theories will Provide More Accurate Predictions. *The British Journal for the Philosophy of Science* **45**, 1-35.

Gillies, D.A. (1973). *An Objective Theory of Probability.* London: Methuen,.

Honderich, T. (1988). *A Theory of Determinism*. Oxford: Clarendon Press.

Howson, C. (1995). Theories of Probability. *The British Journal for the Philosophy of Science* **46**, 1-32.

Hume, D. ([1739] 1888). *A Treatise of Human Nature*. Edited by L.A. Selby-Bigge. Oxford: Clarendon Press.

Hunt, G.M.K. (1987). Determinism, Predictability and Chaos. *Analysis* **47**, 129-133.

Laplace, P.S. (1819). *Essai philosophique sur les probabilités*. In: Oeuvres, vol. 7. Théorie analitique des probabilités. 3eme edition, 1820.

Magee, B. (1971). *Modern British Philosophy*. London: Secker and Warburg.

Mackie, J.L. (1978). Failures in Criticism: Popper and his Commentators, *The British Journal for the Philosophy of Science* **29**, 363-375.

Mellor, D.H. (1971). *The Matter of Chance*. Cambridge: Cambridge University Press.

Meyerson, E. (1930). *Identity and Reality*. London: Allen and Unwin.

Miller, D.W. (1994). *Critical Rationalism: A Restatement and Defence*. Chicago and La Salle: Open Court.

Miller, D.W. (1996). The Status of Determinism in an Uncontrollable World. In: Paul Weingartner & Gerhard Schurz (eds.), *Law and Prediction in the Light of Chaos Research*, pp. 103-114. Berlin: Springer Verlag.

Miller, D.W. (2006). *Out of Error*. Aldershot: Ashgate Publishing.

Miller, D.W. (forthcomming). Popper's Contribution to the Theory of Probability and Its Interpretation. In: Jeremy Shearmur & Geoffrey Stones (eds.), *The Cambridge Companion to Popper*. Cambridge: Cambridge University Press.

O'Hear, A. (1980). *Karl Popper*. London: Routledge.

O'Hear, A. (1985). *Mind* **94**, 453-471.

Popper, K.R. (1934). *Logik der Forschung*. Vienna: Springer.

Popper, K.R. (1957). *The Poverty of Historicism*. London: Routledge and Kegan Paul. Originally published in *Economica* in 1944-1945.

Popper, K.R. (1959). *The Logic of Scientific Discovery*. London: Hutchinson.

Popper, K.R. (1963). *Conjectures and Refutations*. London: Routledge and Kegan Paul.

Popper, K.R. ([1966] 1972). *Of Clouds and Clocks*. St Louis Missouri: Washington University. Reprinted in Popper (1972).

Popper, K.R. (1972). *Objective Knowledge*. Oxford: Oxford University Press.

Popper, K.R. (1973). Indeterminism Is not Enough. *Encounter* **40**, 20-26. Reprinted in Popper (1982a).

Popper, K.R. (1982a). *The Open Universe*. London: Hutchinson.

Popper, K.R. (1982b). *Quantum Theory and the Schism in Physics*. London: Hutchinson.

Popper, K.R. (1990). *A World of Propensities*. Bristol: Thoemmes Antiquarian Books Ltd.

Russell, B.A.W. (1918). On the Notion of Cause. In: *Mysticism and Logic*. London: Allen and Unwin. Reprint: Melbourne, London & Baltimore: Pelican Books, 1953.

Stewart, I. (1989). *Does God Play Dice?* Oxford: Blackwell.

Watkins, J. (1974). The Unity of Popper's Thought. In: P.A. Schilpp (ed.), *The Philosophy of Karl Popper*. (The Library Of Living Philosophers, vol. 14), pp. 371-412. La Salle, IL: Open Court.

Watkins, J. (1984). *Science and Scepticism*. Hutchinson, London.

Watkins, J. (1985). Second Thoughts on Landé's Blade. *Journal of Indian Council of Philosophical Research* **2**, 13-14.

Alan Musgrave

CRITICAL RATIONALISM

1. Introduction

They say that family quarrels are the worst of all. That certainly seems true of family quarrels among philosophers. In several publications (see my 1989a, 1991, 1993a and 1993b) I have tried to say what I think is distinctive about Popper's critical rationalism, including his solution to the problem of induction. My efforts have drawn only sharp criticism from "Popperians" – though not, I am pleased to say, from Popper himself. Ian Jarvie objected strenuously in letters. I had a row with John Watkins (see my 1989b and Watkins 1991). David Miller also took strong exception to my views (see his 1994). The situation is odd, indeed. It contributes to the odd situation regarding Popper's philosophy itself. The self-styled "Popperians" think it is the bees' knees – the philosophical community at large either rejects it out of hand or ignores it. Some scientists endorse it – all philosophers of science think it fatally flawed. Popper and the "Popperians" constantly extol the virtues of reason – the world thinks Popper is a sceptic and irrationalist.

What is the reason for the odd situation regarding Popper's philosophy? No doubt there are several reasons. I think the most important intellectual reason is that Popper's chief contribution to philosophy has still not been understood. This is so despite the fact that Popper has devoted thousands of words to explaining what that contribution is. It has not been understood by the "Popperians," whose efforts on its behalf convince no-one but themselves, and who, as I say, have objected strenuously to what I have said on the matter. Perhaps it was not even understood by Popper himself. (This paradox is, by the way, consistent with Popper's own view that any work has "a life of its

In: E. Suárez-Iñiguez (ed.), *The Power of Argumentation* (*Poznań Studies in the Philosophy of the Sciences and the Humanities*, vol. 93), pp. 171-211. Amsterdam/New York, NY: Rodopi, 2007.

own" and may not be fully undertood by its producer.) What follows is highly controversial. In vainglorious moments I think that I am the only person who understands Popper, including Popper himself. More cautiously, what I should say is this: either Popper answers Hume in the way I think, or he has no answer and his numerous critics on the point are right.

That there is every reason to revisit these matters was shown by the public reaction to Popper's death on September 17, 1994. The problem of induction and its central importance were apparent. Even John Watkins, in a glowing obituary in *The Times* of London for 19 September, after mentioning Popper's claim to have solved Hume's problem of induction, felt compelled to add "This claim has been much questioned." In a scandalous piece in *The Independent* for 19 September, Rom Harré trots out the standard objection, to be discussed in due course, and concludes that Popper's "claim to have solved the problem of induction must now be rejected." Well, let us see.

2. The Problem of Induction

What exactly is Hume's problem of induction? As Popper sees it, the problem is posed by the following argument of Hume's:

> We do, and must, reason inductively.
> Inductive reasoning is logically invalid.
> To reason in a logically invalid way is unreasonable.
> Therefore, we are, and must be, unreasonable.

The problem posed by this argument of Hume's is: "can Hume's irrationalist conclusion be avoided?". This way of putting the problem is due to Popper. Other ways of putting the problem are more common: can induction be justified?; even, how can induction be justified? But these ways of putting the problem beg the question against some possible ways of answering Hume, notably Popper's.

An exegetical aside. The argument just attributed to Hume is not, of course, in Hume's own words. Hume's discussion is couched in the terminology of "impressions" and "ideas." It is also intricately bound up with his discussion of causality. However, the contention is that we can set these complications aside and that when we do we get the argument given above. The contention is, in other words, that the argument given above departs from the letter, but not from the spirit, of Hume's own presentation. That contention might be disputed. But I want to avoid exegetical issues and focus on more important ones. So if you think the stated argument does not capture the essence of Hume's own argument, call it a "Humean" argument instead, an argument prompted by reading (perhaps *misreading*) Hume.

Popper's formulation of the problem highlights Hume's originality. The invalidity of inductive reasoning had already been pointed out by Sextus Empiricus, Francis Bacon, and countless others. It was well-known that the fact that all experienced swans were white does not ensure that the next one will be, let alone that they all are. Hume's originality was to combine this logical triviality with other premises to show that any belief which transcends the immediate evidence of the senses is unreasonable. What other premises? First, that the invalidity of induction means that all experienced swans having been white does not establish that the next one will be white or that they all are, nor does it establish that the next one will *probably* be white, will more likely be white than not, or that they all are. Second, that it is only reasonable to believe what we can establish to be true or probable. This argument runs:

> *Inductive scepticism*: Induction is invalid: no evidence-transcending belief can be established as true or probable.
> *Justificationism*: It is reasonable to believe something if and only if it has been established as true or probable.
> *Irrationalism*: Therefore, no evidence-transcending belief is reasonable.

A further exegetical aside. Some argue that Hume did not even consider, let alone attack, "probabilistic induction" got by prefacing the conclusions of inductive arguments with "probably." Perhaps. But Hume's point that inductive arguments cannot establish truth naturally extends to the further point that they cannot establish probable truth either. Call this natural extension "Humean," rather than "Hume's," so as to avoid further exegetical issues.

3. Non-Solutions

Hume's conclusion was "No evidence-transcending belief is reasonable." Anyone who disagrees must say "Some evidence-transcending beliefs are reasonable." Here I assume that if one person says *P*, anyone who disagrees must be saying or implying "It is not the case that *P*." This assumption can be disputed. Consider the assertion that the Holy Spirit proceeds from the Father alone, and not from the Son also (*filioque*). The *Filioque* dispute divides the Roman Catholic from the Orthodox Churches of the East and has led to wars and to the deaths of countless people. What is a rational person to do with that dispute? Should a rational person disagree and say that it is not the case that the Holy Spirit proceeds from the Father alone and not from the Son also? This, while it is not wrong, seems to be entering into the dispute. No, a better response is to *change the subject.*

Perhaps a better response to Hume's "No evidence-transcending belief is reasonable" is also to change the subject. Perhaps we should forget about beliefs and their reasonableness or otherwise, and focus on the *contents* of those beliefs, the theories, propositions, or whatever, that are believed. Perhaps the rational response is to fly to Popper's and Frege's "Third World."

I doubt it. Popper says that, like E. M. Forster, he "does not believe in belief." What Forster meant is that we should not have unreasonable beliefs, faiths, dogmatic commitments. This is an eminently sensible piece of (second-world) advice. Critical rationalists are not the only people who accept it. And anybody who accepts it thinks there is a distinction between reasonable beliefs and unreasonable ones. At which point Hume's shocking conclusion that no evidence-transcending belief is reasonable confronts us once again.

One way of changing the subject, or of *seeming* to change the subject, is to eschew the term 'belief' in favour of other terms. We are not to believe evidence-transcending hypotheses, but rather choose them or prefer them (when the aim of inquiry is truth) or classify them as true. Terminological legislation like this does not help. To believe something is to think it true. To choose or prefer something when your cognitive aim is truth is also to think it true, that is, believe it. And to classify something as true is also to think it true, that is, believe it. Terminological legislation like this sits oddly with those who think, with Popper, that "words do not matter." Besides, choosing and preferring and classifying are all something that epistemic subjects do, are all "second world" phenomena.

You really *do* change the subject if, browbeaten by Humean scepticism and irrationalism, you give up on truth as the aim of inquiry and put something weaker in its place, whether it be the ability to solve puzzles (Kuhn), the ability to solve problems (Laudan), empirical adequacy (van Fraassen), or empirical adequacy as far as we have looked into the matter (Watkins). These proposals eschew belief in one respect, endorse it in another. We are not to believe or think true any evidence-transcending proposition. We are to believe or think true meta-claims to the effect that this or that evidence-transcending proposition is a good puzzle or problem solver, is empirically adequate, or is empirically adequate as for as we have looked into the matter. Of course, changing the subject in this way does not answer Humean scepticism and irrationalism. Rather, it endorses them and moves on to other topics.

Another way of changing the subject, while not appearing to do so, is to retain belief in the truth of evidence-transcending propositions *but to go in for a peculiar concept of truth itself.* Thus there is talk of an "empirical adequacy theory of truth," according to which an empirically adequate theory is (by definition) a true theory. Such a conception of truth is hopeless, for reasons which need not now concern us. Nor does it help. To believe a theory, or think it true, turns out to be, when the meaning of the word "true" is cashed out, to think true a meta-claim to the effect that the theory is empirically adequate. A similar objection can be levelled at any of the "epistemic" conceptions of truth that are popular nowadays.

(It is a nice question whether the various truth-substitutes or deviant truth-concepts succumb to *analogues* of Humean scepticism and irrationalism about truth. I have pursued this elsewhere and will not pursue it here.)

The scholasticisms discussed in this section were a digression. The point of the digression was to show that Humean scepticism and irrationalism are to be taken seriously. They are not to be evaded by confining oneself to "third world considerations," or by striving ineffectually to avoid the term "belief," or by changing the subject by dropping truth, or by going in for some peculiar epistemic conception of truth. They are to be confronted head-on.

4. Popper' Solution A: Induction Is a Myth

Popper confronts them head on. Popper grapples directly with Hume's argument, which was:

> We do, and must, reason inductively.
> Inductive reasoning is logically invalid.
> To reason in a logically invalid way is unreasonable.
> Therefore, we are, and must be, unreasonable.

Other philosophers who have tangled with this argument accept its first premise and avoid its conclusion by rejecting either the second premise, or the third, or both. Hence the innumerable attempts to "validate" or "justify" induction. Popper disputes Hume's first premise. He says that induction or inductive reasoning is a myth.

It is important to realize that Popper is not here making a factual claim. He is not saying (or he ought not to be saying) that as a matter of fact no one has ever argued from the fact that all observed *A*s were *B*s to the prediction that the next one will be or to the generalization that they all are. That factual claim seems false. It seems false of scientists, who are inveterate generalizers. Kepler figures out that Mars goes in an ellipse, and immediately concludes that all the planets go in ellipses. Mendel does a few experiments on pea-plants, and immediately concludes that his results apply to all pea-plants, nay, to all plants, nay, to all sexually-reproducing plants and animals. It seems false of ordinary folk in the common affairs of life, who are also inveterate generalizers. Small children do not repeat the experiment of touching a hot radiator to see if it still burns them. It seems false of cats and dogs as well. Remember Popper's own story of the generalizing puppy, who did not like the smell of the first and only cigarette it sniffed and thereafter ran from anything looking remotely like a cigarette.

Some will object that generalizing from experience like this is not reasoning inductively. Rather, it is jumping to conclusions. Inductive reasoning requires that the observed instances must be sufficiently large in number. Inductivist orthodoxy backs this objection. But inductivist orthodoxy is problematic: *how* large must the number of observed instances be, to turn rash jumping to conclusions into sober inductive inference? Besides, inductivist orthodoxy is not really available to Popper. As he sees it, inductive reasoning *is* just jumping to conclusions, whether the premises concern a thousand cases, or a hundred, or just one.

The matter is complicated by the fact that some scientific theories cannot be got by inductive generalization, however broadly we construe that scheme of inference. I refer to those theories that transcend the evidence not just in generality, but also in precision and/or observability. Generalizing from imprecise observation cannot get you to a precise theory or to a theory that postulates unobservable entities and processes, as Duhem was the first to see. But this complication aside, it seems that scientists and ordinary folk and animals all do engage in inductive reasoning.

I say that this *seems* to be so. And I was similarly cautious a couple of paragraphs back. The reason for the caution is that reasoners seldom, if ever, state all of the premises they are assuming. We usually, perhaps always, have to reconstruct the arguments being employed. Deductivism is the view that deductive logic is the only logic that we have or need. Deductivists can always reconstruct what look like non-deductive or inductive arguments as deductive arguments with missing premises of one kind or another. I have argued elsewhere (in 1989a, 1989c and 1988) that it conduces to clarity if we do treat them so. For now I merely note that, if accepted, it would enable us to make good Popper's claim that induction is a myth.

Suppose that this claim is wrong. In other words, suppose that induction is not a myth, that scientists and ordinary folk and animals all do engage in inductive reasoning. Would Popper's answer to Hume be empirically refuted by this? It would not. For it is Hume's assumption that we *must* reason inductively that is Popper's real target. This assumption is universally endorsed by philosophers (except Popper). Everyone (except Popper) assumes that we must reason inductively *if we are to have any justified beliefs that trancend the evidence of the senses.* And everyone (except Popper) assumes that we must have justified evidence-transcending beliefs *if we are to have reasonable evidence-transcending beliefs.* We must look deeper.

5. Popper's Solution B: Abandon Justificationism

As we saw, Hume's real originality in all of this was not to point out the invalidity of induction. Rather, it was to come up with an argument that we pedantically reconstructed thus:

> *Inductive scepticism:* Induction is invalid: no evidence-transcending belief can be established as true or probable.
> *Justificationism:* It is reasonable to believe something if and only if it has been established as true or probable.
> *Irrationalism:* Therefore, no evidence-transcending belief is reasonable.

This argument is valid. So if we reject its conclusion, we must reject (at least one of) its premises. Popper accepts the first premise and rejects the second. Popper rejects justificationism. Popper thinks it reasonable to believe things despite the fact that they have not been established either as true or as probable.

That Popper does think this is obscured by two things. It is obscured by Popper's preference for terminological variants of the term 'believe', such as 'prefer [with respect to truth]'. The following passage is typical (1976, p. 104):

> Although we cannot justify a theory . . . we can sometimes justify *our preference* for one theory over another; for example if its degree of corroboration is greater.

The second smokescreen is verisimilitude. Popper stopped asking whether this theory is true and that one false, and started asking whether this theory is closer to the truth than that one. Notoriously, Popper's definition of versimilitude is no good and other definitions are controversial. More important, for our purposes, replacing truth with verisimilitude cuts no *epistemic* ice. If we can never rationally adopt an evidence-transcending hypothesis as true, then neither can we ever rationally adopt an evidence-transcending hypothesis as close to the truth. Popper was wrong when he said that "the search for verisimilitude is a clearer and more realistic aim than the search for truth" (1972, p. 57). Let us leave these obfuscating smokescreens behind, and return to justificationism.

The world thinks justificationism obvious and Popper's rejection of it crazy. To believe something is to think it true. So a reason for believing something must show that what is believed is true or at least probable, more likely to be true than false. Or so Watkins, for example, thinks. He says that justificationism is "underwritten by the truism that to justify the meta-statement '*T* is true' is *eo ipso* to justify *T* itself" (1991, p. 358). But Watkins' truism is irrelevant and question-begging. The relevant meta-statement is not "*T* is *true*" but rather "It is reasonable to believe *T*." And the relevant question is whether to justify or give a reason for believing *T* is *eo ipso* to justify or give a reason for *T* itself.

The term "belief" is notoriously ambiguous. It can refer to the thing believed or the content of the belief, and it can refer to the mental act or state of believing that thing or content. (Here a metaphysical question intrudes. What are belief-contents? Are they abstract entities like propositions or can they be accounted for as non-abstract entities? Let me set this question aside. Let me assume that two people can believe the same thing without saying what that thing is.) Talk of the reasonableness or otherwise of a belief inherits this ambiguity. Is it the belief-content which is reasonable or otherwise, or is it some person's believing that content?

Obviously, in the first instance anyway, it is the latter. One person might reasonably believe something and another unreasonably believe precisely the same thing. Construing belief-contents as propositions for the moment (though nothing essential will hinge on this), it seems like a category mistake to describe a proposition as "reasonable" or "unreasonable." When we say "That's a reasonable proposition," what we mean is that it was reasonable for the speaker to have asserted the proposition, and by implication to have believed it also. So it seems, in the first instance anyway, that what is reasonable or unreasonable are our believings of things, not the things we believe.

Traditionally, however, attention is immediately directed onto the things we believe, the belief-contents. Traditionally, it is taken for granted that a reason for our believing something must be a reason for the something that we believe. That reason might be a *conclusive* reason, which establishes that what we believe is true. Or it might be an *inconclusive* reason, which only establishes that what we believe is more likely true than false. (Or, perhaps, establishes that what we believe is more probable than we previously thought – I shall for the most part ignore this curlicue). No category mistake is committed by the following traditional principle:

J   *A*'s believing that *P* is reasonable if and only if *A* can justify *P*, that is, give a conclusive or inconclusive reason for *P*, that is, establish that *P* is true or probable.

After all, to believe something is to think it true. So if you reasonably believe something you must have a conclusive or inconclusive reason for what you believe, you must have something which establishes that what you believe is true or probable. Or so the world thinks.

Principle J says much the same as the second premise, labelled "*Justificationism*," of the Humean argument repeated at the start of this section. I said that Popper rejects the second premise of Hume's argument. Popper also rejects Principle J. And he is right to do so. For what is it to "have a conclusive or inconclusive reason for what you believe, [to] have something which establishes that what you believe is true or probable"? In the case of conclusive reasons, is it to be aware of a deductively valid argument whose conclusion is the thing you believe? Nothing is easier. The argument "*P*, therefore *P*" is as conclusive and rigorously valid as any argument is. Other deductively valid arguments for *P* merely go around the houses and beg the question in less obvious ways. Nor does it help to say that being aware of the argument is not enough, that you must also be aware that the premise(s) of the argument are true. If you are aware that *P* is true, then the argument "*P*, therefore *P*" is no use to you. Other deductively valid arguments for *P*, whose premises are stronger than *P*, are even less use to you.

The situation with inconclusive reasons is less clear. But there are (*modulo* your theory of logical probability) always propositions in the light of which other propositions are more probable than not (or more probable than they were before you considered those propositions). Those propositions – the "evidence propositions" – had better be known to be true, of course. Let us suppose that they are. Even so, Miller and Popper have an argument that it does not help. Let the hypothesis (belief-content) we are interested in be h, and the evidence (assumed to be known) e. Hypothesis h is equivalent to the conjunction "$(h \lor e)$ and $(e \supset h)$," where the first conjunct is the common content of e and h, and the second conjunct expresses the excess content of h over e. The evidence e entails the first conjunct, and is not a good reason for it (by the preceding paragraph). And given plausible assumptions, the evidence e will *undermine* the second conjunct, in that p $(e \supset h, e)$ will be less than p $(e \supset h)$. This is an ingenious argument. If it is sound, it shows that the whole business of seeking reasons, conclusive or inconclusive, for belief-contents or propositions is fatally flawed. I do not know whether the argument is sound. It has spawned quite a literature, which I cannot enter into here.

But quite apart from this ingenious argument, looking for reasons for propositions is foolish. It is foolish because looking for conclusive reasons for propositions is foolish – and conclusive reasons for propositions would be, were they to be had, the best reasons we could have. The probabilistic stuff about inconclusive reasons, and the literature that it has spawned, do not undermine this simple point.

Popper's rejection of Hume's justificationist premise, or equivalently of Principle J, enables him to endorse Hume's inductive scepticism while rejecting Hume's irrationalism. It also means that what are reasonable or unreasonable are our believings of things, rather than the things that we believe. It is acts of belief (actions of believing?) that are reasonable or rational, not the things we believe, belief-contents, propositions, theories, or whatever. There are precedents for this view. Pascal's wager gives a reason for believing that God exists which is not a reason for God's existence. The pragmatic vindication of induction is reason for believing that Nature is Uniform which is not a reason for the Uniformity of Nature. These precedents are unsavoury ones, of course. I mention them, not to support Popper's central idea, but to help explain what that idea is. Let us be clear about one thing, however: *the theory of rationality is, of its essence, a theory about the "second world," a psychologistic theory.*

Rejecting justificationism is only the negative part of Popper's achievement. To deny Hume's irrationalism is only to deny that no evidence-transcending belief is reasonable. We now need a positive story about which evidence transcending beliefs are reasonable. Popper has a positive proposal.

6. Critical Rationalism CR

Critical rationalism's central contention that if a hypothesis has withstood our best efforts to show that it is false, then this is a good reason to believe it *but not a good reason for the hypothesis itself.* If we have a good reason to believe something, then our belief is a reasonable or rational belief. In other words, provided some hypotheses do withstand criticism the total irrationalist view that all beliefs are unreasonable is mistaken, as is the Humean irrationalist view that all evidence-transcending beliefs are unreasonable. A wedge has been driven between scepticism and irrationalism.

Here a few curlicues need to be mentioned and a few promissory notes issued. There are different kinds of hypotheses and different ways of criticizing them. We need an account of all this. We also need an account of which kinds of criticism are serious criticisms and which not. We need, in short, a *theory of criticism.* Existing critical rationalist literature goes a good way to provide this. An especially important point is that *sceptical criticism* of a hypothesis, devoted to showing that it has not been established as true or even probable, is not criticism of the hypothesis at all.

Suppose we have a theory of criticism. It is important to recognize its limitations. In the critical rationalist model of inquiry we typically (perhaps always) operate with competing hypotheses concerning the matter in hand, whatever that might be. If two competing hypotheses equally well withstand criticism, the critical rationalist will accept neither of them. Here the proper attitude is skeptical "suspension of judgment," supplemented by the hope that future critical activity will enable us rationally to choose. But it is no part of critical rationalism to suppose that we will be able to have a rational belief regarding all contentious issues.

Here I seem to be in disagreement with John Watkins. Watkins replaces truth with "possible truth" as part of the aim of science – the other part is "depth." "Possible truth" turns out to mean not just logical consistency but also consistency with available evidence. Watkins considers a schematic case where T_1 and T_2 are competing incompatible theories, with T_2 deeper and wider than T_1. So far T_1 and T_2 are equally well-corroborated. Watkins counts it a virtue of his position that he can prefer T_2, the deeper and wider theory, as better fulfilling his "optimum aim for science." I on the other hand, must either say that we have no cognitive reason to prefer T_2 to T_1 (the austere view) or that "we do have a reason, namely its being deeper and wider, to adopt T_2 *as true,* even though this reason is *not* a reason for regarding T_2 as true [the hypocritical view]" (1991, p. 356).

Well, the "hypocritical view" is a nonsense: there is no difference between preferring a theory with respect to truth, adopting a theory as true, and regarding a theory as true. I take the "austere view." We "suspend judgment" between T_1 and T_2. That does not stop us judging that T_2 is deeper and wider than T_1 ,and as such would be preferable *if it were true.* What of the claimed superiority of Watkins ability to prefer T_2 as better fulfilling his "optimum aim for science"? It is entirely spurious. When that "optimum aim" is spelled out, what Watkins' preference comes to is: (a) the two theories are equally well corroborated; (b) T_2 is deeper and wider than T_1, and as such would be preferable *if it were true.*

Watkins attacks me with another schematic scenario (1991, p. 357). Here we have three theories. The first has been refuted. The second has been less well corroborated than the third. We agree that the first, refuted theory should be rejected. For the rest my view is said to make a "baleful" difference. But Watkins does not understand my view. He says, first (1991, p. 357):

> Assume, to begin with and as a non-inductivist should, that a theory's being the best corroborated in its field does not give us a reason to regard it as true . . . the Musgrave amendment would oblige one to withdraw into a state of paralyzed indecision: it would make a difference, and a baleful one.

But Watkins' assumption here is not my own, and he knows it. So this first point is irrelevant. Watkins' second point (1991, p. 357) is that my actual view:

> . . . would allow one to say that, although its being corroborated gives us no reason to *regard* T_3 as true it does give us a reason *to adopt* it as true, which seems less coherent than my view.

A mark of scholasticism is the making of distinctions without a difference. Here Watkins distinguishes, marking the distinction with italics, *regarding* something as true and *adopting* something as true. The italics just give us more terminological variants on *believing* something. The key issue, which Watkins never addresses, is whether the non-existence of reasons for what we believe entails the non-existence of reasons for our believing those things.

Let us leave these scholasticisms and return to the heart of critical rationalism, summarized by the following principle:

CR   It is reasonable to believe that P (at time t) if and only if P is that hypothesis which has (at time t) best withstood serious criticism.

Some persist in thinking that to point out that a hypothesis has best withstood serious criticism is to attempt to justify that hypothesis, to show that it is true or probable. Others object that the fact that a hypothesis has best withstood serious criticism does not establish that it is true or probable. Both parties miss the main point. The main point is the rejection of justificationism. This is what enables CR to specify a reason for believing P which is neither a conclusive nor an inconclusive reason for P itself. As critical rationalists see things, our epistemic predicament is that there are no conclusive or inconclusive reasons for our hypotheses. Induction being invalid, as Hume said, there are no conclusive reasons for them. The probabilistic programme having foundered, there are no inconclusive reasons for them either. If we combine these sceptical results with justificationism, the result is irrationalism. But if we reject justificationism a new possibility opens up. There is nothing more rational than a thorough and searching critical discussion. Such a discussion may provide us with the best reason there is for believing (tentatively) that a hypothesis is true – though not, of course, with a conclusive or inconclusive reason for that hypothesis.

CR means, to give an everyday example of the kind Hume discussed, that if bread nourished us when we ate it from Monday through Saturday, then it is reasonable to believe that bread nourishes and to deduce the prediction that it will nourish tomorrow (Sunday). CR means, to give a scientific example of the kind Popper discussed, that if one of our competing evidence-transcending scientific theories best withstands criticism then it is reasonable for scientists to believe that theory and to use it in practical applications.

Of course, the predictions we make from reasonably believed hypotheses might always be mistaken. Sunday's bread could poison us. The next prediction of even the best tested and corroborated scientific theory might turn out to be false. That this is so is a simple consequence of the invalidity of inductive reasoning, which was correctly pointed out by Hume. When a prediction is falsified we will say that what we predicted was wrong, not that it was wrong (unreasonable) of us to have predicted it.

Which leads us to another objection. CR means that we might reasonably believe something false! Is that acceptable? Well, of course it is. Any reasonable theory of reasonable belief must make room for reasonable beliefs in falsehoods. If all the evidence and argument points towards some hypothesis, CR says it is reasonable to believe that hypothesis. If the hypothesis subsequently turns out to be false, we will say that what we believed was wrong, not that it was wrong (unreasonable) of us to have believed it. All that the ever-present possibility of error shows, for critical rationalists, is that our beliefs should always be tentative and that we should always be receptive to new criticisms of them. (As we will see, the same goes for the ever-present possibility of *logical* error: just as we may reasonably believe a falsehood, so also we may even reasonably believe a *logical* falsehood!)

7. Does CR Smuggle in Induction – A?

Can it be so simple? The most widespread criticism of Popper's philosophy is that induction must be smuggled in somewhere. But the countless people who have urged this objection, in one form or another, *themselves smuggle in assumptions which Popper rejected.*

To investigate this, let us confine ourselves (as the critics usually do) to Popper's notion of the high degree of corroboration that a severely tested but unrefuted hypothesis may enjoy. (This will involve an obvious modification to CR, restricting it to corroboration.) Popper says that it is reasonable to adopt such a hypothesis as true (to believe it). Popper also says that it is reasonable to use that hypothesis to make predictions. Here is the Achilles Heel where induction must be smuggled in! Popper must be assuming that predictions from well-corroborated hypotheses will be true, while predictions from refuted hypotheses will be false. Or he must be assuming that predictions from well-corroborated hypotheses will be more likely to be true (more probable) than predictions from refuted hypotheses. Either way we have an inductive principle. Some tried to answer Hume and validate induction by invoking metaphysical principles like "Nature is uniform." Hume said this would not do. Popper says he agrees with Hume. But Popper must be invoking a principle of the same kind. He must be assuming that corroboration is a guide to truth, or near-truth, or high probability. But his "degrees of corroboration" were supposed to be backward-looking, they were supposed to report *past* predictive success and failure. Backward-looking reports say nothing about future performance. So Popper must be smuggling in a metaphysical principle linking past success to future success. Either that, or he must endorse Humean irrationalism regarding all evidence-transcending beliefs.

This is the objection urged by dozens of philosophers, stated as forcefully as I am able to state it. Let us now examine it. The objection assumes that it is reasonable for us to believe something if, and only if, we can justify what we believe, that is, show it to be true or probable. This assumption lies behind the claim that Popper "*must* be assuming that corroboration is a guide to truth, or near-truth, or high probability." But Popper has *rejected* the view that it is reasonable for us to believe something if, and only if, we can justify what we believe, that is, show it to be true or probable. Popper agrees with Hume that we cannot show that evidence-transcending beliefs are true or probable. Popper thinks, however, that we may reasonably believe what we cannot justify. Therefore the standard objection begs the question.

To see this more clearly, let us ask whether any inductive argument is involved in adopting a well-corroborated hypothesis as true. The answer is NO. The argument involved, pedantically set forth, is a straightforward deductive argument:

> CR*: It is reasonable to adopt as true (to believe) the best-corroborated hypothesis.
> *H* is the best-corroborated hypothesis.
> Therefore, it is reasonable to adopt as true (to believe) *H*.

But Popper also says that it is reasonable to believe the predictions which well-corroborated hypotheses entail. Surely an inductive argument is involved here! Well, again NO. A (hopefully uncontroversial) principle of applied deductive logic (label it "ADL") is that if it is reasonable to believe *H*, and if *H* entails prediction *p*, then it is also reasonable to predict *p*. Principle ADL figures in another straightforward deductive argument:

> It is reasonable to adopt as true (to believe) *H*.
> *H* entails prediction *p*.
> ADL: If it is reasonable to adopt as true (to believe) *H*, and if *H* entails prediction *p*, then it is reasonable to predict *p*.
> Therefore, it is reasonable to predict *p*.

Here, of course, the first premise is the conclusion of the first deductive argument. So it seems that only valid deductive arguments are involved in adopting evidence-transcending hypotheses and making predictions from them.

Let us be clear about the status of these deductive arguments. I am not claiming that people *acquire* reasonable beliefs in evidence-transcending hypotheses by running through arguments of the first kind. Nor am I claiming that people *make predictions* by running through arguments of the second kind. People adopt evidence-transcending hypotheses as true by surveying the state of the critical discussion. People make predictions by deriving them from evidence-transcending hypotheses they have adopted as true. These two arguments merely set out pedantically what a critical rationalist might say in answer to queries about the reasonableness of a prediction or belief. And, to repeat, they are valid deductive arguments.

Does any inductive *principle* figure among the premises of these two deductive arguments? The only possible candidate is CR*. Now CR* (and CR) are quite different from what are traditionally labelled "inductive principles." Traditional inductive principles were *metaphysical*, CR* (and CR) are *epistemic*. Why were traditional inductive principles metaphysical? Because you need a metaphysical principle to show truth or probability, and because tradition demands (via J) that you need to show truth or probability to show reasonableness. Because Popper has rejected J, he can adopt an epistemic principle while rejecting any metaphysical one.

But is not CR* (or CR) still an *inductive* principle? Let us not quarrel about words. Let us agree to call it an *"epistemic* inductive principle."

Such principles are ubiquitous. Let total irrationalism be the view that it is unreasonable to believe anything (or, for that matter, to accept anything as X for any X other than truth). Nobody is a total irrationalist. Which is to say that everybody accepts some epistemic inductive principle saying when it is reasonable to believe something (or to accept it as X for some X other than truth). Hume accepted the spartan principle "It is reasonable to believe the immediate evidence of the senses and what can be deduced therefrom." (For Hume, of course, the "immediate evidence of the senses" did not concern, say, white swans but rather "impressions" and "ideas" of white swans in our minds. But this is a complication we can ignore.) Probabilists accept the principle "It is reasonable to believe the evidence and anything that is more probable than not given the evidence." It is this principle which lies behind the probabilist slogan that "Probability is the rational guide of life." Explanationists, believers in "inference to the best explanation," accept the principle "It is reasonable to believe the best available adequate explanation of any body of facts."

Even those who, browbeaten by sceptical arguments, give up on truth as an aim of inquiry and put something weaker in its place (see Section 3) have epistemic principles. Van Fraassen seems committed to something like "It is reasonable to accept as empirically adequate the best-corroborated hypothesis." Or perhaps van Fraassen prefers "It is reasonable to accept as empirically adequate the best available adequate explanation of any body of facts." Watkins prefers "It is reasonable to accept as consistent with available data the deepest theory that is consistent with available data."

Epistemic inductive principles are "ampliative" – they enable you to obtain conclusions that do not follow from the other premises of the deductive arguments in which they figure. For example, CR* was not a redundant premise of the argument-scheme in which it figured. To be ampliative in this sense is simply to be non-analytic. Deductivists, like Popper and me, think there are no valid ampliative or inductive or non-deductive arguments, but there are ampliative epistemic inductive principles. And anyone who is not a total irrationalist (that is, everyone) accepts at least one of them.

Return to the standard objection to Popper. It was said that because a corroboration report is "backward looking" it must be supplemented with some metaphysical inductive principle if it is to have any "forward looking" implications. Of course a corroboration report is "backward looking": one cannot report the results of *future* tests. A corroboration report was a premise of the first deductive scheme:

> CR*: It is reasonable to adopt as true (to believe) the best-corroborated hypothesis.
> *H* is the best-corroborated hypothesis.
> Therefore, it is reasonable to adopt as true (to believe) *H*.

The assertion that it is reasonable to believe a prediction was the conclusion of the second deductive scheme:

> It is reasonable to adopt as true (to believe) *H*.
> H entails prediction *p*.
> ADL: If it is reasonable to adopt as true (to believe) *H*, and if *H* entails prediction *p*, then it is reasonable to predict *p*.
> Therefore, it is reasonable to predict *p*.

No metaphysical inductive principle appears among the premises of either scheme. We need no such principle to forge a "past-future" link. General hypotheses do that job for us – one they are (rationally) adopted as true. So much for the usual objection to Popper's solution to Hume's problem.

8. Does CR Smuggle in Induction – B?

Clearly, Popper's solution stands or falls with CR* or with the more general CR. It may be objected that CR is an arbitrary proposal, that no reason has been given for it, that it is completely unjustified. After all, it has been admitted that CR is ampliative or non-analytic. So we *need* a reason for it. But any reason or justification for CR will be a metaphysical inductive principle of the kind Popper is supposed to be avoiding. *Why* does Popper think it reasonable to believe well-corroborated hypotheses, or more generally, hypotheses that have best stood up to criticism? It must be because he thinks corroboration (or mere generally, having survived criticism) is a good guide to truth or high probability. A metaphysical inductive principle must be invoked to justify CR. Induction is smuggled in after all!

This objection has been described as "unanswerable" (Watkins 1991, p. 358). Let us see. The demand for a *reason for* CR assumes for epistemology precisely what Popper's critical rationalist epistemology denies. It assumes that it is only reasonable to accept CR if a reason can be given *for* CR. A consistent critical rationalist should refuse to give a reason for CR. As the objection rightly states, such a reason would be a metaphysical inductive principle, anathema to critical rationalists, and no use anyway as we would immediately be asked for a reason for the reason, and so on *ad infinitum*.

But this does not mean that CR must be adopted arbitrarily. A consistent critical rationalist can give a reason for *the adoption of* CR. A consistent critical rationalist can argue that it is reasonable to adopt CR because it has withstood criticism better than its rival justificationist epistemic principles. Popper argues this way when he describes our epistemic predicament as he sees it. There just are no good reasons or justifications for hypotheses that transcend the evidence. Inductive arguments being invalid, as Hume said, there are no conclusive reasons for them. The probabilist programme for answering Hume having foundered, there are no inconclusive reasons for them either. Combine these sceptical results with the justificationist principle that it is only reasonable to believe what we can justify, and it will follow that no evidence-transcending belief will be reasonable. But reject justificationism and a new possibility opens up. The best we can do is provide reasons for believing evidence-transcending hypotheses which are not reasons for those hypotheses themselves. There is nothing more rational than a thorough and searching critical discussion. Hence the best reason we can have for believing an evidence-transcending hypothesis is that it has best survived such a discussion. This is the reason for adopting CR.

Some of this, such as the rejection of probabilism, is still highly contentious. And even if it is accepted that CR withstands criticism better than rival epistemic principles (a big "if"), another objection immediately presents itself. All this is *circular*! The critical rationalist is saying that it is reasonable to adopt CR *by CR's own standard of when it is reasonable to adopt something*! The critical rationalist is arguing:

> CR: It is reasonable to adopt the theory which best survives critical scrutiny.
>
> CR best survives critical scrutiny.
>
> It is reasonable to adopt CR.

Nobody sceptical of CR is going to be convinced that it is reasonable to adopt it by an argument that *assumes* it.

Is this objection devastating? I submit that it is not. For what are the alternatives to a circular argument of this kind? One alternative is to give no reason for adopting CR, to admit that your belief in your theory of reasonable belief is unreasonable. Bartley accused Popper of admitting this, and Popper accepted the accusation. (I do not think he should have, but that is another story.) The only other alternative is to give a reason for adopting CR which is not of a critical rationalist kind, again admitting that your belief in your theory of reasonable belief is unreasonable by your own lights, and inviting a reason for your reason, and so on *ad infinitum*. Circularity is preferable to either of these alternatives – just.

Here I simply agree with William Warren Bartley III. He argued (in his 1962 and 1964) for the superiority of a comprehensively critical rationalism, a rationalism that could subsume itself and be rational by its own lights. A less than comprehensive rationalism, Bartley urged, would be subject to the *tu quoque* objection "But you too are also irrational – about your theory of rationality." Warming to the theme, Robert Nozick argued that the circularity involved in self-subsumptive theories is not a vice but a virtue (1981, p. 138):

> It seems plausible that philosophy should seek to uncover the deepest truths, to find . . . justificatory principles so deep that nothing else yields them, yet deep enough to subsume themselves. Reaching these should be a goal of philosophy, so when that situation occurs with some topic or area, instead of a crisis we should announce a triumph.

I wish this were right. But it is not. Self-subsumption is too easy to obtain. "It is reasonable to believe anything said in a paper by Alan Musgrave" subsumes itself, since it occurs in this paper, but it is a crazy epistemic principle. So is "Granny told me I ought to believe everything she tells me." And "The Pope declared *ex cathedra* that everything declared *ex cathedra* by the Pope is a matter of faith" is no triumph either.

No, self-subsumption or circularity is not a virtue – it is merely, at this level of abstraction, the least of the vices. Count this as an objection to critical rationalism, if you like. But take comfort from this – *any general theory of reasonable belief will be subject to the same objection.* Any general epistemic principle is either acceptable by its own lights (circularity), acceptable by other lights (hence irrational by its own lights and inviting an infinite regress), or not rationally acceptable at all (irrational again). So even though the rational adoption of CR involves circularity, this cannot be used to discriminate against it and in favour of some rival theory of rationality. An objection that hits all competitors with equal force is no objection to any particular one. Consider a parallel: it is no objection to any particular evidence-transcending hypothesis that it has not been proved, since the same applies to all competing evidence-transcending hypotheses.

9. Miller's Mess

The foregoing (especially Section 5) owed much to David Miller, particularly his "A Critique of Good Reasons," which has become Chapter 3 of his (1994). Yet Miller and I are in sharp disagreement. In this section I try to untangle the mess.

Miller's thesis is that "there are no such things as good reasons" (1994, p. 52). Does he mean reasons for believings or reasons for beliefs? It turns out that he means both. He means both because he ignores the distinction between belief-acts and belief-contents. And he ignores that distinction because he implicitly *accepts* justificationism, which was, you will recall, the following principle:

J   *A*'s believing that *P* is reasonable if and only if *A* can justify *P*, that is, give a conclusive or inconclusive reason for *P*, that is, establish that *P* is true or probable.

This leads Miller to attempt to square the circle – to defend *reason* while insisting that there are no good *reasons* for anything. But if there are no good reasons for beliefs, then there are no reasonable beliefs. Critical rationalism as Miller understands it (and as Popper understands it if Miller has Popper right) is wholesale irrationalism in disguise. You do not answer Hume by agreeing with him.

Miller's adoption of J is almost explicit when he says: "For scarcely anyone these days thinks that any factual hypothesis can be established beyond doubt, or that sufficient reasons can be given for *thinking* any hypothesis to be true . . . " (1994, p. 58, italics mine). That Miller implicitly adopts J is obvious. His arguments are those rehearsed in Section 5. Those arguments are entirely devoted to showing that there are no good reasons for belief-contents, propositions, theories, hypotheses, or whatever. It follows that nobody can *have* a good reason for a belief-content. It does *not* follow that nobody can have a good reason *for believing* some belief-content. Only if you implicitly adopt J will you think that once you have established that there are no good reasons for belief-contents you have also established that there are no good reasons for believings. Yet Miller slides effortlesssly from the former to the latter.

In documenting this slide, I must first note a terminological fad. Miller prefers to talk, not of beliefs, but of the acceptance or rejection of hypotheses, and of classifying hypotheses as true or false. As I already said (in Section 3), nothing is changed by this terminological fad. To accept a hypothesis as true is to believe it (and to accept it tentatively is to believe it tentatively). To classify a hypothesis as true is also to believe it (and to classify it tentatively is to believe it tentatively).

The slide starts early. Miller announces that:

> . . . there are no such things as good reasons; that is, sufficient reasons for *accepting* an hypothesis rather than rejecting it, or *rejecting* it rather than accepting it . . . (1994, p. 52, italics mine)

Here what is denied are good reasons for accepting or rejecting things (that is, for believing things). In the next sentence but one we have:

> Yet the illusion persists that the rational person is the person who can supply good reasons *in favour of what he thinks* . . . (1994, p. 52, italics mine)

Miller announces three theses, the first of which is that "it is impossible to furnish a good reason in favour of any thesis whatever" (1994, p. 55). The second thesis is that good reasons for any thesis are redundant, the third that they are unnecessary. Notice how like all global sceptical theses, these theses self-destruct: if there is no good reason for any thesis whatever, then there is no good reason for these theses either. Miller sees the point and says he is not going to argue for his theses (1994, p. 55).

But of course, he does argue for them. The main argument is that canvassed in Section 5. And to repeat, that argument entirely concerns belief contents or propositions. Yet here are some of the claims made about belief-acts or preferrings or (as Miller prefers) classifyings as true:

> . . . no reason at all is needed for classifying a statement as true . . . (1994, p. 64)

> . . . we need no reason for classifying h as false . . . (1994, p. 70)

> . . . we should be prepared to classify statements as true or false without having a reason for so classifying them. (1994, p. 71)

> . . . the critical rationalist's answer to the question "Why do you think that h is true?" . . . will be "Why not?" (1994, p. 71.)

This is, I submit, not critical rationalism but wholesale irrationalism. Miller describes our epistemic task as that of sorting truth from falsehood. Critical rationalism's basic idea is that the best way to go about this is to subject our hypotheses to the severest criticisms we can devise (including where appropriate severe experimental tests), and classify those that survive the ordeal as true and those that do not as false. Both classifications are tentative and fallible, of course. But surely the claim is that it is *reasonable* to proceed in this way. Let the case be a clear-cut one. Let there be just two available hypotheses h_1 and h_2, and let h_1 be as soundly refuted as a hypothesis can be and h_2 as strikingly corroborated as a hypothesis can be. Surely the claim is that the fact that h_1 has been soundly refuted is a reason to classify it as false, while the fact that h_2 has been strikingly corroborated is a reason to classify it as true. Suppose a person flies in the face of the critical discussion and classifies h_1 as true and h_2 as false. Given his professed views, Miller cannot say that these classifyings are unreasonable, while those of the critical rationalist are reasonable.

I think that Miller overlooks the key feature of critical rationalism, its rejection of justificationism (that is, of principle J) and its positive proposal that if a hypothesis has withstood our best efforts to show that it is false, then this is a good reason to believe it *but not a good reason for the hypothesis itself.* Of course, anyone who thinks Principle J self-evident will think that Miller is *right* to ignore these things. If Principle J is self-evident, critical rationalism is a peculiar position indeed. How can there be good reasons for believings (classifyings-as-true, preferrings-with-respect-to-truth) which are not good reasons for the things believed (classified, preferred)? Well, as I understand it critical rationalism asserts, rightly or wrongly, that our epistemic predicament is such that the former reasons are all that are to be had. My complaint is that Miller ignores this possibility. Tacitly and therefore uncritically assuming Principle J, he argues against good reasons for the things we believe (or classify as true), and concludes without further ado that there are no good reasons for our believing those things (our classifying them as true). The result is irrationalism, despite the rationalist rhetoric.

In his book, Miller ignores my criticism of his position and attacks my own position. He complains that I do not divulge what use can be made of claims to the reasonableness of beliefs (1994, p. 107). Goodness me, the epistemic problem of problems is "What should I believe?". We are contending with Hume's claim that any evidence-transcending belief is unreasonable. I should have thought it perfectly obvious what use such claims are. They enable us to solve the epistemic problem of problems. And they enable us to answer Hume instead of changing the subject. Miller has a "metamathematical parallel" which I find incredible. We are interested in whether a mathematical theory is consistent. Painstaking efforts to show that it is not meet with failure. Miller says that this failure "plainly deserves to be reported" (1994, p. 108). (I wonder why.) He continues: "But to report as an additional item of news that we have found a good reason to believe in the theory's consistency seems to me to add nothing but words" (1994, p. 108). Suppose one person believes that the theory is consistent, and another believes that it is inconsistent. Are these two beliefs equally reasonable, or rather equally unreasonable? Miller, it would seem, thinks they are.

Miller later returns to the attack on what he calls my "cosmetic rationalism." He talks of my "facile little derivation":

> It is reasonable to accept theories that have survived severe critical scrutiny.
> H has survived severe critical scrutiny.
> Therefore, it is reasonable to accept H.

Labelling the major premise of this argument P, he complains that "To be in a position to conclude that it is reasonable to accept H, we would still need to know that P is true" (1994, p. 123). No, say I. If we believe A, and A implies B, then we are "in a position to conclude B." And if our belief in A is a reasonable belief, so is our belief in B. Later Miller asks what we are to do with the conclusion of my "facile little derivation," saying "We are patently not mandated to accept H, or classify it as true . . . There is a gap between the acceptability of the theory H and the theory H itself that Musgrave's variety of critical rationalism is powerless to span" (1994, p. 124).

This is to misunderstand the *purpose* of my "facile little derivation."
After surveying the critical discussion, we have accepted *H*, classified it
as true, come to believe it. If somebody queries the rationality of this, the
"facile little derivation" can be produced. When we believe *A*, we might
then use *A*, rather than "I believe that *A*," as a premise to get further
conclusions. Miller prefers us to talk of "accepting *H* in a free act (or
conjecture)" (1994, p. 124). Apparently, when we do this we are allowed
to use *H* as a premise in further arguments. Why the asymmetry? Why
does Miller not insist that our premise can only be "I freely conjecture
that *H*," a premise which, being a statement about me, will have none of
the (non-trivial) consequences that *H* does?

The further complications in Miller's argument, which have to do
with my contention that Principle CR is rationally acceptable by its own
lights, are vitiated by the same problem. I shall not enter into those
complications.

What of Miller's own position? It is, to be blunt, a crazy position. If a
theory is testable then we "admit it into science" and "classify it as true."
We then test it. If it survives tests, we continue to classify it as true. If
not, we classify it as false. Why the latter? Why is it that " . . . we
should . . . discard falsified hypotheses . . . " (1994, pp. 79-80)? Why is it
"rationally indefensible to adopt or hold a criticizable position that has
not survived severe criticism" (1994, p. 80)? Must it not be that the
failure of a position to withstand severe criticism is a *good reason* not to
adopt or hold that position? What else might "rationally indefensible"
mean? But according to Miller there are no good reasons for anything. As
for admitting things into science and classifying them as true, this is
madder still. Let two inconsistent hypotheses be admitted into science, as
often happens. We violate the law of contradiction if we classify both of
them as true.

10. Bartley's Comprehensively Critical Rationalism CCR

Earlier I agreed with Bartley that a comprehensive or self-subsumptive
rationalism is best. But critical rationalism as I understand it,
encapsulated in CR, is probably to be distinguished from Bartley's
Comprehensively Critical Rationalism CCR. Sometimes Bartley seemed
to be saying that a belief is reasonable if what is believed is *criticizable*.
But this is a crazy thesis. It means that it is reasonable for us to believe
that the Moon is made of green cheese, since this view is eminently
criticizable – and indeed has been thoroughly *criticized*. Critical

rationalism as I understand it is made of sterner stuff. According to CR it is thoroughly *unreasonable* for us to believe that the moon is made of green cheese, *because* that view has been thoroughly criticized and refuted.

But I probably misunderstand CCR. The exegetical problems here are considerable. Much of the time Bartley seems to be saying that a belief is reasonable if the believer takes a certain *attitude* or *stance* towards it, that of "not being irrationally committed to it" or "holding it open to criticism" or "being willing to listen to criticism and take it seriously." Now I know what it is to hold a belief. But I am not sure that I know what it is to "hold a belief open to criticism." It cannot be anything as strong as to *accept* any criticism once it is offered and to reject the belief, since criticisms can themselves be criticized and rejected. Anything weaker threatens to be too weak. CCR, viewed as a particular attitude or stance or cast of mind, is in dire need of further explanation. Whatever further explanation might be given, it is clear that it is a thoroughly "second world" or "psychologistic" theory.

Sometimes the position seems different again. It is not that a belief is reasonable if a certain attitude is taken towards it. Rather, the position is that only that attitude is properly described as reasonable, not the beliefs it is directed at.

As I said, Bartley might not have thought that it is reasonable to believe anything that is criticizable. Indeed, Miller complains that this is a "caricature of CCR" (1994, p. 92). But this is a rum business. Bartley did sometimes write as if CCR is just CR. Miller quotes the following, for example (1994, p. 79):

> A position may be held rationally without needing justification at all – *provided that it can be and is held open to criticism and survives severe examination.*

Now I assume that to "hold a position" is to adopt it as true or believe it. And I assume that to hold a position rationally is to have a reason for adopting it as true or believing it. Miller tries to convince us, however, that for Bartley "it is the method of investigation, not its outcome, that is rational (or not rational)" (1994, p. 79). Miller concedes that the quoted passage seems to conflict with his own view that a hypothesis may be held rationally "even if it has just this moment been conjectured" (1994, p. 80). He then tries to make out that Bartley agrees with him after all. I do not think he succeeds in this. But as I said, the exegetical problems are formidable – and intrinsically unimportant.

If Bartley's CCR is what I call CR, well and good. If it is a caricature to say that according to CCR anything criticizable is reasonably believed, well and good again. My "caricature" is no straw position – my "caricature" is *precisely Miller's own position*!

11. Critical Rationalism and Logical Omniscience

ADL said:

> ADL: If it is reasonable to adopt as true (to believe) *H*, and if *H* entails prediction *p*, then it is reasonable to predict *p*.

Here is an objection to ADL, proposed by Greg Currie. Let it be reasonable by critical rationalist lights to adopt *H* as true. Let *H* be not merely false but contradictory, though nobody knows this. Then ADL tells us that it may be reasonable to adopt as true not just a falsehood but a *logical* falsehood, a statement of the form "*P* and it is not the case that *P*" which follows from *H*. That cannot be right. It can never be reasonable to believe a *contradiction*!

Currie thinks it obvious that it can never be reasonable to believe a contradiction. How naïve! There are logicians – Australia is full of them – who think that some contradictions are *true*. It is called "dialethic logic." If a contradiction may be true, then presumably it may sometimes be reasonable to believe a contradiction. But I am no fan of dialethic logic. If a contradiction may be true, then a *reductio ad absurdum* argument loses its force. And if philosophers cannot invoke *reductio ad absurdum* arguments, then they (and others) lose their chief critical weapon. I once asked Graham Priest, the high priest of dialethic logic, what became of *reductio* arguments on his view. He said it was a good question. He answered that "*Reductio* is not always valid." Logicians used to think an argument-form was valid, full-stop, or invalid, full-stop, not valid here and invalid there. But logic, like other things, is (as David Stove might have said, and as Pavel Tichy did say) in its "Jazz Age." The motto of our philosophical times comes from Irving Berlin:

> In olden days a glimpse of stocking
> Was looked on as something shocking,
> Now Heaven knows,
> Anything goes!

So, I do not block Currie's objection by saying that some contradictions are true.

I first thought to block it by modifying ADL so that it requires that *H* is *known* to entail *p*. My thought was that a known contradictory consequence *p* would be recognized as such, rejected, and *H* rejected with it. But that is no good. If *H* is contradictory then it entails a contradiction, whether or not anyone knows this. The trouble, if there is one, lies with CR rather than ADL. The trouble, if it is one, is that CR seems to license reasonable belief in a contradictory theory. Suppose prolonged and serious attempts to show that a theory is contradictory all fail. Suppose, in other words, that the hypothesis that the theory is not contradictory has survived serious criticism. Then CR licenses reasonable belief that the theory is not contradictory. *The theory may be contradictory for all that*!

Is this a problem? I think not. Those who think otherwise think that reasonable folk are logically omniscient. But we should not think this. A contradiction may be deep and difficult to discover. Here is a made-up story to illustrate the point.

I am a passenger on an ocean liner. Up the gang-plank I go, to see a sign saying "SHIP'S BARBER: I am the man who shaves every man on the ship who does not shave himself." I acquire the reasonable belief that there is a ship's barber, who shaves every man on the ship who does not shave himself. After all, official signs on ships, airplanes, and the like are generally right. (There will be more on why this belief is reasonable in Section 14.) Now, Bertrand Russell is a fellow-passenger: he tells me that it cannot be. For who shaves the ship's barber? If the ship's barber shaves himself, then he is not shaved by the ship's barber. If the ship's barber does not shave himself, then he is shaved by the ship's barber. Russell shows, in other words, that my reasonable belief that there is a ship's barber (a man who shaves every man on the ship who does not shave himself) is contradictory.

Of course, after Russell has explained this to me it is no longer reasonable for me to believe that there is a ship's barber – assuming, of course, that I understand and accept Russell's argument. Producing or understanding an argument *takes time*. What was reasonably believed at time *t'* may not be reasonably believed at a later time *t'*, after new evidence or argument has come in. In this case the new "evidence" is *logical* evidence.

The qualification 'assuming that I understand and accept Russell's argument' in this story is a necessary one. I might have been so far from logical omniscience that I could not understand Russell's argument. In which case I might have persisted in my reasonable – though logically false – belief that there is a ship's barber. A person who is *very* far from logical omniscience is the Tortoise in Lewis Carroll's delightful *What the Tortoise Said to Achilles* (1894). Achilles presents the Tortoise with a very simple valid deductive argument (it does not matter which one) from premises A and B to conclusion Z. The Tortoise says that he accepts A and B, but not Z. We do not know whether the Tortoise rejects Z, that is, accepts "It is not the case that Z," or whether he is agnostic about Z, that is, neither accepts it nor rejects it. In the former case, which is admittedly unlikely, the Tortoise has a contradictory set of beliefs. Since A and B logically imply Z, the Tortoise's beliefs entail both Z and "It is not the case that Z." But this is *not* to say that the Tortoise believes both Z and "It is not the case that Z." The Tortoise does not believe Z at all. The Tortoise does not see the validity of Achilles' simple argument and does not actually infer Z from A and B.

To think otherwise is to think that people believe all the logical consequences of the things they believe, to think that people are logically omniscient. There are plenty of good reasons not to think this. For two thousand years people thought certain very simple Aristotelian arguments valid which George Boole thought invalid. It is plausible to suppose that the human brain can store only finitely many beliefs, whereas any belief has infinitely many consequences. Nor must those infinitely many consequences be *trivial* ones, as when B has as consequences "Either B or C," "Either B or D," and so on. Let B combined with some unthought-of C entail an unthought-of D. Then B by itself entails the conditional "If C then D." There are infinitely many such conditionals. And many of them will be non-trivial. When Newton devised his theory, he had no idea that it would come to be applied to the motions of double stars.

Bill Bartley has said all this before. He writes (1987, p. 436):

> . . . *we never know what we are talking about* . . .When we produce and affirm a theory, we also propose its logical implications. (Otherwise we should not have to retract it when these come to grief.) That is, we affirm all those statements that follow from it – as well as further implications that result from combining this theory with other theories that we also propose or assume. But this means that the informative content of any idea includes an *infinity of unforeseeable* nontrivial statements.

But the first statement, the one in italics, seems to *assume* logical omniscience. It seems to assume that we know what we believe (or "affirm") only if we know all the consequences of what we believe (or "affirm"). But we don't. We know what we believe all right. Could we believe that P without knowing what P is? We can know what it is that we believe without knowing or believing all the consequences of what we believe.

Two paragraphs back the Tortoise had a contradictory set of beliefs but did not, because of his logical short-sightedness, believe a contradiction. What should Achilles have done with the Tortoise? Obviously, he should have tried to convince the Tortoise of the validity of his argument from A and B to Z. He should have explained that (1) an argument is valid if it is impossible for the premises to be true and the conclusion false. He should then have pointed out that (2) it is impossible for A and B to be true and Z false. And he should then have concluded that (3) the argument from A and B to Z is valid. But this is to produce another argument, a "meta-argument" from (1) and (2) to (3). This second argument is actually more complicated than the one poor Achilles started with! Suppose the Tortoise is not just logically short-sighted but logically *blind*. He might well say that he accepts (1) and (2) but not (3).

What Achilles actually does in Carroll's dialogue is produce a statement which is tantamount to (2). He says "If A and B are true, Z must be true." Once we recognize this for the meta-statement that it is, and once we remove the misconditionalisation which puts "must" in the wrong place, what we have is "It must be the case that the conditional statement 'If A and B, then Z' is true." And this statement is logically equivalent to (2): the argument from A and B to Z is valid. What does the Tortoise do? He insists that Achilles writes "If A and B are true, Z must be true" as an extra premise (it is labelled C), and says he accepts A, B, and C, but not Z. An obvious infinite regress ensues, involving ever more complicated arguments with ever more complicated meta-statements as redundant premises.

What is Carroll's message? Characteristically he does not spell it out. But at one point Achilles says that logic will take the Tortoise by the throat and *force* him to accept Z. Carroll's message is that Achilles is wrong. Logic cannot take a logically blind person "by the throat." To put it another way, you cannot teach logic to a logically blind person, a person who has no "logical intuition" whatever, a person who will not simply see or accept the validity of some simple arguments. (In my colleague Pavel Tichy's way of teaching first-order logic, as a pure system of natural deduction, the only argument-form that students need to simply see or accept as valid is *very* simple. It is this:

The argument-form $P_1, \ldots, P_n \therefore P_i$ (where i ranges from 1 to n) is valid.

But of course, a logically *blind* person is not going to accept even this.

The message runs deeper still. Suppose people *were* logically omniscient. Suppose that as a matter of psychological fact that as soon as people come to believe *P* they cannot help also coming to believe any *C* which is a consequence of *P*. Then we could view the laws of logic as "laws of thought," laws which describe the way people cannot help thinking. Logic would become a branch of psychology. This is, in fact, how logic was viewed earlier in the nineteenth century. We have only to think of the title of George Boole's major work – *The Laws of Thought*. Though to be fair to Boole he did come to see that the laws of logic do not so much *describe* how people do and must think as *prescribe* how they ought to think (see my 1972). Notoriously, Frege and Husserl explicitly attacked psychologism. Their main argument against it was that the prime business of logic was to sort out valid from invalid reasoning, and that no mere description of how people reason could do this, since people reason invalidly as well as validly. The deeper message of Carroll's dialogue reinforces this criticism.

The twentieth-century has seen a revival of psychologism. The crudest version of modern psychologism is the "inference-license view of laws and/or generalizations" defended by Ryle, Toulmin, Harré and Hanson. It is an irony of history that all of these invoke Carroll's dialogue in support of their view! But I have criticized all that elsewhere (see my 1980) and will not repeat myself here.

12. Experience, Perceptual Belief, and Principle E

I have talked of evidence, of "evidence-transcending hypotheses" and of "all the evidence pointing towards some hypothesis." What is *evidence*? What role does *evidence* play in the critical rationalist story?

The role of evidence is obvious enough. The best way to criticize scientific hypotheses is to pitch them against the deliverancies of observation and experiment. But are these deliverancies of observation and experiment yet more hypotheses to be pitched against the further deliverancies of observation and experiment?

Well, sometimes they are. I seem to see a book. But I suspect hallucination, so I reach out and try to touch it, too. This is primitive testing of the hypothesis that there is a book before me. If I do touch the book as well, then CR licenses belief that there is a book before me.

Still, it is a fanciful scenario. It might have been less fanciful had my example been drawn from scientific experimentation rather than everyday observation. But let us not obscure the fact that most of the time we do not treat our perceptual beliefs this way. Only when we have some specific reason to suspect perceptual error do we "check out" a perceptual belief. Most of the time we simply acquire perceptual beliefs from perceptual experiences and then use them to "check out" other beliefs that we might possess. Is a critical rationalist to deem "unchecked" or uncriticized perceptual beliefs unreasonable beliefs because they have not withstood criticism? That would be odd. It would mean that in the process of empirical testing we would be assessing the reasonableness of nonperceptual beliefs by seeing how well they fit with unreasonable perceptual beliefs!

Critical rationalism needs supplementing, therefore, with a concession to the epistemic primacy of sense-experience. And it is clear from the foregoing what that concession needs to be (I label it "E" for "Experience"):

E   It is reasonable to perceptually believe that P (at time t) if and only if P has not failed to withstand criticism (at time t).

Clearly, the earlier principle CR will require modification so as to restrict it to non-perceptual belief:

CR   It is reasonable to non-perceptually believe that P (at time t) if and only if P is that hypothesis which has (at time t) best withstood serious criticism.

Principle E introduces an asymmetry into the epistemic situation. A non-perceptual belief is reasonable if it has best withstood criticism – a perceptual belief is reasonable if it has not failed to withstand criticism. The latter is just the commonsense view "Trust your senses unless you have a specific reason not to."

Principle E is a concession to the epistemic primacy of sense-experience. But it is a concession of quite a different kind than the traditional empiricist one. It is not that perceptual beliefs are all true and certain. It is not even that perceptual beliefs are all more likely to be true than false. A perceptual belief may be subjected to criticism just like any other belief and found wanting. The contention is merely that unless and until this happens a perceptual belief is a reasonable belief. Experience is not a source of certain or even probable knowledge – it is merely a source of reasonable (reasonably adopted) belief.

The introduction of Principle E means that we are owed an account of the difference between perceptual and non-perceptual belief. The distinction will be partly logical: (the contents of) perceptual beliefs will be particular rather than general, since the senses inform us about goings-on in the world at particular times and places. The distinction will also be partly causal: perceptual beliefs are brought about by sensory experiences and these involve causal transactions with the world.

But the perceptual beliefs we form do not depend solely on the way the world irritates our sensory nerve endings. They also depend on the conceptual or linguistic resources that we possess. To use a time-worn example, a person who lacks the concept of a galvanometer (a person who does not know what the word 'galvanometer' means) cannot form the perceptual belief that there is a galvanometer on the table as a result of seeing the galvanometer on the table. Quite so. (And by the way, the trendy notion that such a person cannot even see the galvanometer is just a confusion between seeing-that and seeing. Suppose that a cat lacks the concept of a galvanometer. Still, puss can see the galvanometer perfectly well, as evidenced by the fact that she does not bump into it when the mouse she is chasing hides under it.) But a person who knows what a galvanometer is *can* form the perceptual belief that there is a galvanometer on the table as a result of seeing it. And such a person can also, by reading the galvanometer, form the perceptual belief that the wire to which it is connected is carrying a current of 15 amps. For the critical rationalist these count as perceptual beliefs despite the highly "theoretical" character of terms like 'galvanometer' or 'current'. For within critical rationalism there is no search for a "theory-neutral observational vocabulary" in which "observation reports" might be formulated which are somehow fully justified by the experiences that prompt them. Nor, I need scarcely add, is there any attempt to secure an infallible empirical basis for science by confining perceptual beliefs to reports about how things currently *appear* to be, let alone to so-called "sense-data reports." Critical rationalism takes a thoroughly *realist* view of perception: the *objects* of perception are things and events in the world, not events in our minds or brains. As a result, our perceptual beliefs are fallible beliefs.

Still, Principle E does introduce an asymmetry, and involves a whiff of what is called "foundationalism." According to Principle E, experience gets us started. Experience provides us with lots of reasonable beliefs against which the reasonableness of other beliefs can be assessed. Which is just to say that critical rationalism is a thoroughly empiricist point of view.

13. An Evolutionary Argument

Principle E can be buttressed by an argument from the theory of evolution. (The argument is hinted at by Popper – see, for example, his 1972, p. 77.) If the theory of evolution is to be believed, our sense-organs are evolutionary products designed (by natural selection) to give us information about the world which will guide our actions in the world. True, the information they give us is highly *selective*. Other critters with different concerns have different sensory systems and get different information from them. But selective information is not *false* information. To think otherwise is to commit the no-truth-but-the-whole-truth fallacy: "John is tall" is false, because it omits to tell us whether or not John is bald. (This fallacy permeates idealist philosophy, incredible though that may appear.) But a sensory system which systematically misled its possessor about (selected) features of the world would, if the theory of evolution is to be believed, have been eliminated by natural selection. Ergo, since the theory of evolution *is* to be believed, it is sensible to trust our senses unless we have a specific reason not to. Which is just Principle E.

The trouble is that this argument is circular. (Alston argues in his 1993 that *all* arguments for the reliability of sense perception are circular in the same way.) If the theory of evolution is to be believed, that is because it best withstands criticism, including criticism from experience. But – the argument runs – the perceptual beliefs which fail to refute the theory of evolution are only reasonably adopted because the theory of evolution is to be believed. One cannot justify Principle E by invoking the theory of evolution. And that should be no surprise: no *proposition* can be justified, and Principle E is a proposition. The question should be whether one can justify the adoption of Principle E by invoking the theory of evolution. The answer is that even this is circular. The most that this argument can show is that the commonsense critical rationalist attitude to experience and the theory of evolution are mutually supportive. This is important. It is important because most epistemology, indeed most philosophy, is pre-Darwinian. Since Berkeley, most philosophers have been idealists of one kind or another. Idealism, which makes reality mind-dependent, conflicts with the Darwinian view that the world existed long before any human minds did.

The argument from evolution, as well as being circular, is controversial. I said that a sensory system which systematically misled its possessor about (selected) features of the world would, if the theory of evolution is to be believed, have been eliminated by natural selection.

This has been denied. Devitt and Sterelny give the example of the excessively timid mouse, which exhibits predator-avoidance behaviour in the absence of predators more often than in the presence of a predator. They say (1987, p. 249):

> Natural selection does not favour true beliefs . . . Thus, it will not matter to the survival of the mouse that it is mostly wrong when it thinks "Look out, a predator." What matters is that it is always right when it thinks "All clear."

The timid mouse is a counterexample only if we assume that the slightest noise produces in the mouse the belief and associated behavioural response "Look out, a predator." What if the slightest noise produces in the mouse the belief/response pair "Look out, a predator-like noise"? Now the mouse's perceptual beliefs are typically true rather than typically false. In this way we might be able to defuse apparent counterexamples by exploiting the deep and difficult problem of content-ascription to inner states.

I have just talked, as Devitt and Sterelny do, of the mouse's perceptual beliefs or thoughts. There are those who think it makes little sense to ascribe beliefs or thoughts to creatures which lack language, as the mouse presumably does. I think they are wrong. But it would take us too far afield to argue the point here.

A different objection (due to my colleague Colin Cheyne) is that the evolution of our sensory system cannot create any presumption in favour of perceptual beliefs involving recently-invented things like television sets or telephones or galvanometers. An earlier example concerned a person who reads a galvanometer and forms the perceptual belief that the wire to which it is connected is carrying a current of 15 amps. A similar example would be acquiring the perceptual belief that Clinton is in Washington from watching the TV news. The acquisition of such beliefs depends as much on other beliefs that the person possesses as it does on sensory stimulations.

There are those who will insist that the real perceptual beliefs in these cases are "The galvanometer is reading 15 amps" and "It says on the TV news that Clinton is in Washington." People then put these perceptual beliefs together with their other beliefs and *infer* that the current in the wire is 15 amps or that Clinton is in Washington. I prefer not to go down this road. It is a road which, consistently traveled, leads to the doctrine of sense-data, to the doctrine that our perceptual beliefs concern goings-on inside our heads from which all else is inferred. For reasons I have explained elsewhere (1993, pp. 274-279), I do not believe in sense-data.

But can I have it both ways? Is there not a tension between a relaxed attitude as to what is to count as a perceptual belief and the idea that our sensory system does not systematically mislead us? I admit the tension. In other words, I admit that the premise of the argument from evolution is dubious when we think of very "indirect" perceptual beliefs like the ones just discussed.

Colin Cheyne likes to tease me with a parallel argument. He writes (1993, p. 10):

> Perhaps induction, though formally invalid, is to be trusted because if it were not reliable then inductive reasoners would have been eliminated.

Clearly I must resist the parallel. And resist it I do. Cheyne's argument assumes that all or most of us are "inductive reasoners." I regard this as highly problematic, for reasons I have already explained (Section 4, above). Besides, so called "inductive reasoning" is known to be unreliable by any person of sense – which is not the case for the process of acquiring perceptual beliefs. People of sense do not always "reason inductively." People of sense do not argue that the more times a coin has landed heads the more likely it is to land heads again. People of sense do not argue that the more times your joke has made a person laugh the more likely it is to raise a laugh the next time you tell it. People of sense do not argue that the more times you have woken up alive and well the more likely it is you will wake up alive and well the next day.

The general structure of cases like this is obvious enough. We do not "reason inductively" in such cases *because of our other beliefs about the cases.* Adherents of inductive logic have to find a way round cases like this and somehow deem them "inductively invalid." Suppose they succeed. Consider how very peculiar inductive logic is. Whether or not an argument is "inductively valid" depends on how the world happens to be. An argument which is "inductively invalid" would become "inductively valid" in a world where re-told jokes worked best. Inductive reasoning is supposed to be the way we find out how the world happens to be. And whether a piece of inductive reasoning is valid depends on wholly contingent facts about the way the world happens to be.

So I reject Cheyne's parallel. It is not obvious that people are "inductive reasoners." And even if they are, there is no such thing as inductive *logic* which would deem some pieces of inductive reasoning valid and others not.

Cheyne's last objection to the evolutionary argument is that it is meant to show that "our perceptual beliefs are reliable: they are more than likely true" (1993, p. 10). He objects that when it comes to perceptual beliefs I have adopted a *reliabilist* view, and have implicitly assumed that it is reasonable to believe anything which arises from a reliable belief-forming process. But I can see no inconsistency in having a reliabilist view of perceptual beliefs and denying reliabilism in general. Perhaps perception is the only reliable belief-forming process. Cheyne objects that to be consistent I must accept that if any other process P is reliable then P also yields reasonable beliefs. I do accept this *conditional*. But that is consistent with denying that its antecedent can be satisfied. Critical rationalists deny that induction is a reliable process. Critical rationalists also deny that the process they commend is reliable – or at least, they *should* deny this if they to avoid the widespread accusation that they smuggle into their theory either inductive reasoning or some metaphysical inductive principle.

Nor, and this is crucial, need critical rationalism be spiced with reliabilism to achieve this happy result. Reliabilism regarding perceptual belief was only invoked as a premise in an argument for the differential treatment accorded to perceptual beliefs. That argument can be criticized on several counts, as we have seen: it is ultimately circular; its premise can be attacked on general biological grounds; even if that general attack can be resisted, its premise seems weak for "indirect" perceptual beliefs. But even if this argument for the differential treatment accorded perceptual beliefs is judged defective *the differential treatment can remain*. And the reasonableness of adopting that differential treatment can be urged on critical rationalist lines: that the proposal to do so withstands criticism better than rival proposals.

14. Personal Knowledge and the Division of Epistemic Labour

In the previous section I assumed that critical rationalism's concession to reliabilism can be limited to the case of perceptual beliefs. In fact, I doubt that it can. Is not memory a reliable belief-producing process? Could the general reliability of memory be defended by an evolutionary argument like that for perception? Is not testimony a reliable process? Could the general reliability of testimony be defended by an argument predicated on a general human predilection for truth-telling?

These questions, particularly those about testimony, are important. Most of our beliefs are acquired from the testimony of others, parents, friends and neighbours, teachers, radio and TV, books, and newspapers. It would be nice if critical rationalists could treat beliefs acquired from the testimony of others as they treat perceptual beliefs, and deem them reasonable beliefs unless they have failed to withstand criticism. For such a view would contrast happily with the quite bizarre views of traditional epistemologies on the point. Traditional empiricists insist that it is only reasonable for me to believe *what I have verified with the evidence of my own senses*. Traditional rationalists insist that it is only reasonable for me to believe *what I have derived from self-evident "first principles" or "axioms."* Both extreme views have the unhappy consequence that most of my beliefs, and most of yours, and most of everybody else's, are unreasonable. Critical rationalism spiced with reliabilism regarding perception, memory and testimony seems more acceptable. But before I investigate this further, we have to tidy up our three critical rationalist principles.

Those three principles were:

E   It is reasonable to perceptually believe that P (at time t) if and only if P has not failed to withstand criticism (at time t).

CR   It is reasonable to non-perceptually believe that P (at time t) if and only if P is that hypothesis which has (at time t) best withstood serious criticism.

ADL   If it is reasonable to adopt as true (to believe) H, and if H entails prediction p, then it is reasonable to predict p.

These principles are couched in the impersonal mode – "It is reasonable to believe . . . " That sufficed for our general discussions so far. But in the end it will not do. It will not do because it invites the obvious question "Reasonable *for whom?*". Suppose I am a logically short-sighted person who reasonably believes H, who does not realize that H entails p, yet who believes p nonetheless for no reason whatever. Do I *reasonably* believe p? It would seem not, contrary to Principle ADL.

I first thought to block this objection (which is due to Colin Cheyne) by going on about the meaning of the word 'predict'. In the imagined scenario, I have not predicted p – I have asserted it or prophesied it. To predict something is to get it from something that you already have. But this is to require that before Principle ADL can be applied I must have (validly) obtained p from H. So we have to abandon the impersonal mode to make this explicit. And for reasons already explained (in Section 11), we also need to index the principle to time. We obtain:

ADL: If A reasonably adopts as true (believes) H (at time t), and A has validly deduced p from H (at time t), then A reasonably adopts as true (believes) p (at time t).

But what if P is something that A has reason to think false, or absurd, or even contradictory? As we already saw (in Section 11), the very act of deducing a contradiction from what we believe (reasonably or not) provides a criticism of our belief, a reason for not believing it. But the problem is much more general than the extreme case of deriving a contradiction suggests. Suppose Alice reasonably believes that all swans are white. She has validly deduced that the next swan Betty sees will be white. But Betty has reported that she has just seen a black swan. And Alice reasonably believes Betty to be a reliable witness. The conditions of ADL have been met, yet Alice does not reasonably believe that Betty has just seen a white swan. The problem is that Alice's belief has failed to withstand serious criticism – Alice is aware of Betty's report, and reasonably believes Betty to be a reliable witness. So we need to add a further clause to rule this out:

ADL: If A reasonably adopts as true (believes) H (at time t), and A has validly deduced p from H (at time t), and p has not failed to withstand serious criticism from A (at time t), then A reasonably adopts as true (believes) p (at time t).

Principles E and CR also need to be indexed to particular people and times, yielding:

E A's perceptual belief that P is reasonable (at time t) if and only if P has not failed to withstand criticism from A (at time t).

CR A's non-perceptual belief that P is reasonable (at time t) if and only if P is that hypothesis which has (at time t) best withstood serious criticism from A.

But now a problem has become apparent regarding Principle CR. Must I have subjected all my non-perceptual beliefs to serious criticism before they count as reasonable beliefs? I believe that there are nine planets. So do you. I have not checked out that belief in any way, seriously or unseriously. And neither, I suspect, have you. Yet you and I *reasonably* believe that there are nine planets. We got our belief from the astronomers, the people we pay to tell us what to believe about the solar system and other matters astronomical. Paying experts to find things out for us is eminently rational. It puts in place a *division of epistemic labour*, which is eminently rational. The trouble is that Principle CR does not recognize this.

Before I discuss this, let me point out that Principle E is not in the same trouble. If I am aware of no criticism of my perceptual belief that P, then my belief is reasonable. The fact that others have falsified *P* does not alter this – unless they tell me, and I believe them.

One way of dealing with the division of epistemic labour has already been mooted. That was to propose an analogue of Principle *E* for testimony – for "testimentory beliefs," beliefs acquired from other people:

T   *A*'s testimentory belief that *P* is reasonable (at time *t*) if and only if *P* has not failed to withstand criticism from *A* (at time *t*).

I said that Principle E is no more than the commonsense view "Trust your sense, unless you have a specific reason not to." The same might be said of Principle T: "Trust what other folk tell you, unless you have a specific reason not to". Reflection on what we actually *do* regarding the testimony of others will reveal that Principle T captures it pretty well.

A friend tells me that she went to the beach yesterday. I have no reason to suppose that she might be deceiving me. Indeed, I know that she is fond of going to the beach. I immediately form the belief that she went to the beach yesterday. Is my belief a reasonable belief? Principle T says that it is.

A stranger tells me that he has just flown into town. I ask what airline he flew on and he answers "None, I flew into town." I do not believe him. What he said is inconsistent with a pretty firm belief I have, and a reasonable one to boot, that people cannot fly unaided. I have checked what he said against my other reasonable beliefs and found it wanting. What he said has failed to withstand criticism from me, and according to Principle T it would not be reasonable for me to believe it.

Principles E and T are pretty commonsensical – and the same might be said of principle CR. The rejection of justificationism separates the question of the reasonableness of a belief from that of the justification of what is believed. As a result, the formar question becomes much more tractable. Beliefs are transistory states, for beliefs change. Reasonable beliefs are transistory states, too. It was reasonable for Aristotle to believe many things (given the state of the critical discussion in his time) which it is no longes reasonable for us to believe (given the state of the critical discussion in our time). Critical rationalism does solve rather than dodge the central problem into a proper perspective and shows that slightly refined common sense is all we need to solve it.

University of Otago
Department of Philosophy
Dunedin 9054
PO Box 56
New Zealand
e-mail: alan.musgrave@otago.ac.nz

REFERENCES

Alston, W.P. (1993). *The Reliability of Sense Perception*. Ithaca and London: Cornell University Press.

Bartley, W.W. III (1962). *The Retreat to Commitment*. London: Chatto and Windus.

Bartley, W.W. III (1964). Rationality versus the Theory of Rationality. In: M. Bunge (ed.), *The Critical Approach to Science and Philosophy*, pp. 3-31. New York: Free Press.

Carroll, L. (1894). What the Tortoise Said to Achilles. *Mind* 4, 278-280. Reprinted in: *The Complete Works of Lewis Carroll* (London: Nonesuch Press, 1939), pp. 1104-1108.

Cheyne, C. (1993). Some Reasonable Criticisms of Critical Rationalism. Unpublished manuscript.

Devitt, M. and K. Sterelny (1987). *Language and Reality: An Introduction to the Philosophy of Language*. Oxford: Basil Blackwell.

Miller, D. (1994). *Critical Rationalism: A Restatement and Defence*. Chicago and La Salle, IL: Open Court.

Musgrave, A. (1972). George Boole and Psychologism. *Scientia* **107**, 593-608.

Musgrave, A. (1980). Wittgensteinian Instrumentalism. *Theoria* **46**, 65-105.

Musgrave, A. (1988). Is there a Logic of Scientific Discovery? *LSE Quarterly* **2-3**, 205-227.

Musgrave, A. (1989a). Deductivism versus Psychologism. In: M.A. Notturno (ed.), *Perspectives on Psychologism*, pp. 315-340. Leiden: E.J. Brill.

Musgrave, A. (1989b). Saving Science from Scepticism. In: F. D'Agostino and I.C. Jarvie (eds.), *Freedom and Rationality: Essays in Honor of John Watkins*, pp. 297-323. Dordrecht/Boston/London: Kluwer Academic Publishers.

Musgrave, A. (1989c). Deductive Heuristics. In: K. Gavroglu et. al (eds.), *Imre Lakatos and Theories of Scientific Change*, pp. 15-32. Dordrecht/Boston/London: Kluwer Academic Publishers.

Musgrave, A. (1991). What Is Critical Rationalism. In: A. Bohnen and A. Musgrave (eds.), *Wege der Vernunft: Festschrift zum siebzigsten Geburtstag von Hans Albert*, pp. 17-30. Tubingen: J.C.B. Mohr (Paul Siebeck).

Musgrave, A. (1993a). *Common Sense, Science and Scepticism*. Cambridge: Cambridge University Press.

Musgrave, A. (1993b). Popper on Induction. *Philosophy of the Social Sciences* **23**, 516-527.

Nozick, R. (1981). *Philosophical Explanations*. Oxford: Clarendon Press.

Popper, K.R. (1972). *Objective Knowledge*. Oxford: Oxford University Press.

Popper, K.R. (1976). *Unended Quest*. Glasgow: Fontana/Collins

Stove, D. (1982). *Popper and After: Four Modern Irrationalists*. Oxford: Pergammon Press.

Watkins, J.W.N. (1991). Scientific Rationality and the Problem of Induction: Responses to Criticisms. *British Journal for the Philosophy of Science* **42**, 343-368.

NOTES ON THE CONTRIBUTORS

Joseph Agassi was born in Jerusalem in 1927 and educated at Hebrew University and London School of Economics and Political Science. He has taught at Hong-Kong University, University of Illinois, Boston University and was a Research Associate in the Center for Advanced Study in the Behavioral Sciences at Stanford University. He is Professor Emeritus both at Tel Aviv University and York University. He is a member of the American Philosophical Association, the Philosophy of Science Association, the American Association of University Professors and Fellow of the American Association for the Advancement of Science, the Royal Society of Canada and World Academy of Art and Science. His books include: *Toward an Historiography of Science* (1963), *The Continuing Revolutions* (1969), *Hong Kong: A Society in Transition* edited with I.C. Jarvie (1969), *Faraday as a Natural Philosophical Anthropology* (1977), *Science and Society* (1981), *Technology: Philosophical and Social Aspects* (1985), *Toward a Rational Philosophical Anthropology* (with Y. Fried), *Rationality: The Critical View* edited with I. C. Jarvie (1987), *Science and Culture* (2003).

Mario Bunge was born in Buenos Aires, Argentina in 1919 and educated at Universidad Nacional de la Plata. He is Frothingham Professor of Logic and Metaphysics at McGill University in Montreal and Head of the Foundation of Philosophy of Science. He is Fellow of the American Association for the Advancement of Science and a member of the Academie Internationale de Philosophie des Sciences, the Institute International de Philosophie, the Academy of Humanism and the Royal Society of Canada. His books include *The Myth of Simplicity* (1963), *Foundation of Physics* (1967), *Philosophy of Physics* (1972), *Teoría y Realidad* (1972), *La Ciencia, su Método y su Filosofía* (1973), *Treatise on Basic Philosophy* 8 volumes (1974-1989), *Mente y Sociedad* (1989), *Finding Philosophy in Social Science* (1996), *Social Science under Debate* (1998), *The Sociology-Philosophy Connection* (1999), *Scientific Realism* (2001), *Emergence and Convergence* (2003). He received the Premio Principe de Asturias in Communication and Humanities, the

Award of Merit of the University of Winsconsin, and he is doctor Honoris Causa of several Latin American universities.

Fred Eidlin was born in Rochester, USA, in 1942 and educated at Darthmouth College, University of Toronto and Institut d'Etudes Politiques de Paris. Since 1990 he is Professor of Political Science at Guelph University. He is a member of the American Political Science Association, The Societé Quebecaise de Science Politique, and the International Political Science Association. His publications include *Logic of Normalization: A Soviet Intervention in Czechoslovakia* (1980), *Constitutional Democracy: Essays in Comparative Politics*, editor (1983), as well as papers "Meaning, Perspective, and the Social Construction of Reality," "Popper and Democratic Theory," "A Popperian Sermon." He is the editor of the newsletter about Popper's philosophy.

I.C. Jarvie was born in South Fields, England, in 1937 and educated at London School of Economics and Political Science. He taught at Hong-Kong University and since 1969 he has been Professor of Philosophy at York University where he is now Distinguished Research Professor. He is a member of the American Philosophical Association, the Royal Institute of Philosophy, the Royal Anthropological Institute and Mind Association. His books include *The Revolution in Anthropology* (1964), *Hong Kong: A Society in Transition* edited with Joseph Agassi (1969), *Movies and Society* (1970), *Concepts and Society* (1972), *The Story of Social Anthropology* (1972), *Functionalism* (1973), *Towards a Sociology of the Cinema* (1974), *Rationality and Relativism, Movies and Social Criticism* (1978), *Philosophy of the Film* (1987), *Rationality: The Critical View* edited with Joseph Agassi (1987) *Hollywood's Overseas Campaign* (1992), *Popper's Open Society after 50 years* (with Sandra Pralong, 1997), *The Republic of Science. The Emergence of Popper's Social View of Science, 1935-1945* (2001).

Bryan Magee was born in London in 1930 and educated at Christ's Hospital and Keble College, Oxford. He was a writer, critic and broadcaster. In 1973 he was elected a Visiting Fellow of All Souls. From 1974 until 1983 he was a Member of Parliament. At present he is an Honorary Fellow of Wolfson College, Oxford University. He has held visiting appointments at Harvard, Yale, California, and Sidney universities. His books include *The New Radicalism* (1962), *The Democratic Revolution* (1964), *Towards 2000* (1965), *Aspects of Wagner*

(1968), *Modern British Philosophy* (1971), *Facing Death* (1977), *Men of Ideas* (1978), *The Philosophy of Schopenhauer* (1983), *Philosophy and the Real World. An Introduction to Karl Popper* (1985), *The Great Philosphers* (1987), *On Blindness* (1995), *Confessions of a Philosopher* (1997), *Wagner and Philosophy* (2002). In 1979 he was awarded the Silver Medal of the Royal Television Society for his work in broadcasting.

David Miller was born in Watford, England, in 1942 and educated at Cambridge University, London School of Economics and Stanford University. He was Karl Popper's assistant from 1965 to 1967. He is Reader in Philosophy at The University of Warwick. He has been Visiting Professor at Minnesota, Pennsylvania, Arizona, São Paulo (Brazil), Autónoma Metropolitana and Universidad Iberoamericana (Mexico), Cordoba (Argentina) universities. He has been Secretary of the British Logic Colloquium, Honorary Treasurer of the British Society for the Philosophy of Science and Chair of the Programme Committee of the Centennial Congress Karl Popper 2002. His publications include *Critical Rationalism. A Restatement and Defence* (1994), *Out of Error* (2006), as well as papers "Popper's Qualitative Theory of Verisimilitude," "A Geometry of Logic," (with Karl Popper) "Why Probabilistic Suport Is Not Inductive," "External Consequence Operations," "Do We Reason When We Think We Reason, or Do We Think?," "Science without Induction." He is also the editor of *Popper Selections* (1985).

Alan Musgrave was born in England in 1940 and educated at University of London. He was a Research Assistant to Karl Popper and is Professor of Philosophy at the University of Otago. He has held visiting appointments at The University of London, Melbourne University and National University of Australia. He is a member of the British Society for the Philosophy of Science, the Philosophy of Science Association, the Australasian Association for Philosophy, the Australasian Association for History, Philosophy and Social Studies of Science. His publications include *Problems in the Philosophy of Science* (1968), *On a Demarcation Dispute* (1968) and *Criticism and the Growth of Knowledge* (1970), the last three edited with Imre Lakatos; *Popper and the Human Science* edited with G. Curie (1985), *Common Sense, Science and Scepticism* (1993), *Realism and Rationalism* (1999), as well as papers and chapters in books: "The T-scheme Plus Epistemic Truth Equals Idealism," "Conceptual Idealism and Stove's Gem," "Metaphysical Realism versus Word-Magic."

Harold P. Sjursen was born in Plainfield, New Jersey, U.S.A. and educated at the New School for Social Research. He is Professor of Philosophy and Director of Global Exchange Programs and Liberal Studies, Head of Humaniteis and Social Sciences at Polytechnic University in New York. He is a member of the American Philosophical Association, the Society for Phenomenology and Existential Philosophy, the Soren Kierkegaard Society, the Society for Philosophy and Technology. He has published papers about Kierkegaard's philosophy and a book on the individual in the public realm.

Enrique Suárez-Iñiguez was born in Mexico City in 1948 and educated at National Autonomous University of Mexico (UNAM), Sorbonne and Cornell University where he received an appointment as Visiting Professor of Government in 1986-1987. He is Professor of Political Science at National Autonomous University of Mexico and a member of the Mexican Academy of Sciences. Since 1985 he is a National Researcher of his country. His publications include *Eurocomunismo* (1978), *Los Intelectuales en México* (1980), *De los Clásicos Políticos* (1993), *Viejos y Nuevos Problemas de las Ciencias Sociales*, editor (1994), *La Fuerza de la Razón. Introducción a la Filosofía de Karl Popper* (1998), *La Felicidad. Una visión a través de los grandes filósofos* (1999), *Cómo hacer la tesis. La solución a un problema* (2000), *Enfoques sobre la democracia*, editor (2003), *Filosofía Política Contemporánea. Popper, Rawls y Nozick* (2005); as well as papers like "The Role of Political Theory in the Teaching of Political Science in Mexico," "Political Science in Mexico in the Cold War and Post-Cold War Context," "Las Ideas Políticas de Platón," "La obra de Rawls."

Ambrosio Velasco Gómez was born in Mexico City in 1954 and educated at National Autonomous University of Mexico (UNAM), Metropolitan Autonomous University (UAM) and University of Minnesota. He is Professor of Philosophy and Dean of the Liberal Arts School (Facultad de Filosofía y Letras) at National Autonomous University of Mexico (UNAM). He was the founding director of the Graduate Program of Philosophy of Science at UNAM and a member of the Mexican Academy of Sciences and of Mexican Philosophical Association. His publications include: *Teoría Política: Historia y Filosofía* (1995), *Resurgimiento de la Teoría Política en el Siglo XX. Filosofía, Historia y Tradición*, compiler (1999), *Tradiciones Naturalistas y Hermenéuticas en la Filosofía de las Ciencias Sociales* (2000), *El Concepto di Herrística les Ciencies y Les Humanitades*, editor

(2000), *La Ténacité de le Politique*, co-editor (2002), *Hermenéuticas y Progreso Certifico*, co-editor (2003), as well as papers as "Universalismo y Relativismo en los Satan Filisoficas el Tradicion," "Toward a Political Philosophy of Science."

John Watkins was born in Surey, England, in 1924 and educated at the University of London and Yale University. He was Professor of Philosophy at the London School of Economics from 1966 until 1989 and an Emeritus Professor since then. In 1961 he was a Heath Visiting Professor at Grinnell College, USA, in 1977 William Evans Visiting Professor at the University of Otago and in 1983-1984 a Visiting Fellow at the National University of Australia. He was a member of Mind Association and the British Society for the Philosophy of Science and he was its President from 1972 until 1975. He was also been co-editor of the *Journal for the Philosophy of Science*. His publications include *Confirmable and Influential Metaphysics* (1958), *Hobbes System of Ideas* (1965), *Freiheit und Entscheidung* (1978), *Science and Scepticism* (1984), as well as papers and chapters in books "Metaphysics and the Advancement of Science," "Non-Inductive Corroboration," "Otto Neurath," "The Popperian Approach to Scientific Knowledge," "Hume, Carnap, and Popper." He was awarded with the Distinguished Service Cross Royal Navy. He died in 1999.

Printed in the United States
by Baker & Taylor Publisher Services